Food Engineering Series

For further volumes:
http://www.springer.com/series/5996

Matteo Alessandro Del Nobile • Amalia Conte

Packaging for Food Preservation

 Springer

Matteo Alessandro Del Nobile
Department of Agricultural Sciences
   Food and Environment (SAFE)
University of Foggia
Foggia, Italy

Amalia Conte
Department of Agricultural Sciences
   Food and Environment (SAFE)
University of Foggia
Foggia, Italy

ISSN 1571-0297
ISBN 978-1-4899-9633-6   ISBN 978-1-4614-7684-9 (eBook)
DOI 10.1007/978-1-4614-7684-9
Springer New York Heidelberg Dordrecht London

Preface

The food industry faces the task of satisfying increasing consumer demands for food that keeps as long as possible while maintaining the required quality. The development of effective scientific and commercial strategies for meeting these goals is not easy. Various technologies and ingredients are used to meet levels of product quality, but it is difficult to test them for the purpose of assessing how food quality will be maintained over the products' intended shelf life. Packaging can play a key role in food product preservation. Therefore, efforts to improve the performances of packaging solutions and preserve food freshness have been spearheaded in diverse fields. Packaging is usually a composite item meeting various needs, and its design is clearly a fundamental part of new products. Considering the importance of packaging in determining product shelf life, the correct approach entails considering, on the same level of importance, product development and its packaging system. This book addresses important issues associated with the nature of packaging and the shelf life characteristics of some important food types. Such information must be organized and made accessible to the target audience.

Three main topics of food packaging are presented and discussed. In particular, a complete overview of mass transport phenomena in polymers intended for food packaging applications is discussed in depth in the first section. With a strong emphasis on principles, this section provides a solid and comprehensive framework for students and practitioners in that it covers the basic concepts of packaging permeation and provides references commonly used to teach packaging. Students will find the first three chapters an excellent base on which to build their understanding of other, more complicated, explanations of the theory of permeation. The second section describes the most relevant approaches to developing eco-friendly active packaging, including recent fabrication methods and technical information on the advantages and limits of techniques, and the underlining systems that could find application in food. The last section surveys how packaging can help prolong shelf life. The strategies are described using different case studies of various food categories. Much of the content relates to the key issues of the microbial and chemical stability of foods and of the sensory changes that occur in foods in storage. The last four chapters of the book carefully examine issues related to how the quality of raw materials, process conditions, the internal environment created by the packaging system, and the external environment in which food is stored come together to influence the changes that occur in food during storage.

We sincerely hope that this book will help researchers and workers in the many fields related to food packaging, to understand the relevant issues and stimulate further insights.

Foggia, Italy   Matteo Alessandro Del Nobile
Foggia, Italy   Amalia Conte

Contents

Part I
Shelf Life Modeling of Packaged Food

The purpose of this section is to review the basic concepts of shelf life (SL) modeling. First, the general approach will be provided. Direct and mechanistic models will be presented and discussed separately. In particular, the elements of a mechanistic model will be analyzed in detail by presenting the main information available in the literature. Models related to package mass transport properties and the process of food degradation will also be reported. The last part of this chapter will be focused on how package mass transport properties and food degradation process equations can be combined to predict food shelf life.

What is SL modeling about? It is generally recognized that a SL model is a useful tool either for predicting or simply calculating the SL of packaged foods. To do these things, first, a quantitative measure of food quality is needed, then a threshold value for it must be set, but what is needed most is a function that relates food quality to storage time. This section will be concerned with finding the relationship between food quality and storage time.

Actually, there are two possible ways to approach this problem. The fastest way to derive a SL model is to directly provide an equation that relates the packaged-food quality to storage time (i.e., direct model). Several types of direct models appear in the literature. Polynomial equation (I.1), exponential equation (I.2), and power law function equation (I.3) reported, in what follows, are a few examples:

$$FQ(t) = a_0 + a_1 \cdot t + a_2 \cdot t^2 + \ldots\ldots + a_n \cdot t^n, \qquad (I.1)$$

$$FQ(t) = a_0 \cdot \exp(-a_1 \cdot t), \qquad (I.2)$$

$$FQ(t) = a_0 \cdot t^{a_1} \qquad (I.3)$$

where FQ(t) is the packaged-food quality, a_i are the fitting constants, and t is the storage time. Generally, the direct model's parameters (i.e., a_i) do not have any particular physical meaning since these models are not based on a specific picture of the phenomena involved in food degradation.

The mechanistic approach is the other way to find the relationship between food quality and storage time (i.e., to derive a SL model). It consists in first identifying all the phenomena involved in the deterioration process that affect the packaged food during storage, then in providing a quantitative description for each of them, and finally in combining all this information into a single set of equations, which are generally differential equations. The phenomena involved in packaged food degradation can be clustered into two main groups: mass transport properties of the package and food deterioration mechanisms. As an example, the equation used to determine the amount of low molecular weight compound exchanged between the inside and outside of a flexible package under steady-state conditions has the following form:

$$J_{SS} = P \cdot \frac{\Delta p}{\ell},\qquad(I.4)$$

where J_{SS} is the steady-state permeant mass flux, P is the permeability coefficient of the packaging film, Δp is the permeant partial pressure across the packaging film, and ℓ is the packaging film thickness. As an example of a packaged-food deterioration mechanism, the extent of the oxidation reaction rate, Ext(t), of dry foods is reported as a function of the extent of oxidation reaction, water vapor partial pressure, $p_W^{in}(t)$, and oxygen partial pressure, $p_{O_2}^{in}(t)$, in the package headspace (Labuza 1971):

$$\frac{dExt(t)}{dt} = mp \cdot \left(Ext + \frac{M_1 + M_2 \times Ext(t)}{\sqrt{\frac{p_W^{in}(t)}{p_W^*}} \times 100} \right) \cdot \left(\frac{p_{O_2}^{in}(t)}{M_3 + M_4 \times p_{O_2}^{in}(t)} \right)\qquad(I.5)$$

where mp is the mass of packaged food, p_W^* is the equilibrium water vapor pressure, and M_i are the model's parameters.

Generally, the aforementioned elements are integrated by means of balance equations, usually mass balance equations, which combine the package mass transport properties with the packaged-food deterioration mechanisms. Actually, the SL model is a set of equations, usually differential equations, composed of relationships describing the package mass transport properties, the food deterioration mechanisms, and the balance equations, where the food quality indices are unknown functions. Continuing with the example of dry foods, the oxygen and water mass balance equation in the package headspace is as follows:

$$\frac{dn_{O_2}^{ins}(t)}{dt} = A \times J_{O_2} - R_{O_2}\qquad(I.6)$$

$$\frac{dn_{H_2O}^{ins}(t)}{dt} = A \times J_{H_2O},\qquad(I.7)$$

where $n_{O_2}^{ins}(t)$ and $n_{H_2O}^{ins}(t)$ are the number of oxygen and water moles inside the package, respectively; J_{O_2} and J_{H_2O} are the oxygen and water mass flux through the package, respectively; A is the surface area of the package; and R_{O_2} is the oxidation rate given by Eq. I.5. In this specific case, the SL model is composed of Eqs. I.5, I.6, and I.7 and has unknown functions such as Ext(t), $p_{H_2O}^{in}(t)$, and $p_{O_2}^{in}(t)$. The former two functions are the packaged-food quality indices. Solving the aforementioned set of differential equations it is possible to find the relationship between food quality [i.e., Ext(t), $p_{H_2O}^{in}(t)$] and storage time.

As expected, advantages and disadvantages are associated with these two approaches. Normally direct models are simple and empirical and are usually used to calculate SL by means of either data interpolation or small-scale extrapolation. The model's parameters are usually obtained through an experimental data fitting procedure. The other approach is generally more complex since mechanistic models are difficult both to derive and to handle. The most important feature of these types of model is that they are generally predictive and can be used for design purposes. It is worth noting that these models are generally derived by giving a quantitative description of each involved phenomenon. Therefore, they also provide insight into each event occurring during storage.

References

Labuza TP (1971) Kinetics of lipid oxidation in foods. CRC Crit Rev Food Technol 2:355–405

Chapter 1
Direct Models for Shelf Life Prediction

1.1 Introduction

It is often necessary to study various packaging strategies to preserve a specific commodity. To this aim, shelf life (SL) tests are run to determine the effectiveness of certain packaging solutions. In these cases, a model that either extrapolates or interpolates the experimental data (e.g., by a simple data fitting) is generally used to calculate the SL and, consequently, the effectiveness of a given packaging strategy. To get an idea of what these types of models are about, two cases will be presented: one where the quality of the packaged food can be described by means of a single quality index, as is the case with many kinds of fresh-cut produce, and another where the quality depends on more than just one quality index, such as dairy products.

1.2 Single Quality Index Models

Predictive microbiology is a useful tool for determining the SL of food products whenever the microbial cell load is the sole packaged-food quality index. Several attempts have been made to develop a predictive model of spoilage growth inside or on the surface of foods as a function of time. These models may be analytical expressions (direct models), such as the Gompertz or the logistic curve (Zwietering et al. 1991) that exhibit the typical sigmoidal trend of a bacterial growth curve, or are sets of ordinary differential equations (mechanistic models) (Baranyi and Roberts 1995).

   The empirical sigmoidlike analytical expressions used in predictive food microbiology are very appealing mainly due to their simplicity. The accuracy in predicting growth depends on the number of parameters used in the sigmoidal model. Modified versions of the Gompertz equation can include three or more parameters to describe the behavior of the bacterial growth curve. For example,

M.A. Del Nobile and A. Conte, *Packaging for Food Preservation*,
Food Engineering Series, DOI 10.1007/978-1-4614-7684-9_1,
© Springer Science+Business Media New York 2013

a modified version of the Gompertz model to describe a bacterial population was proposed by Zwietering et al. (1990). The kinetic parameters derived by the Gompertz equation have often been used to calculate the SL of numerous minimally processed vegetables (Corbo et al. 2004; Lanciotti et al. 1999; Riva et al. 2001; Sinigaglia et al. 2003). In fact, the SL of these products was calculated by setting the maximum acceptable contamination level to $5 \cdot 10^7$ CFU/g, as determined by French regulations (Ministere de l'Economie des Finances et du Budget 1988). The method adopted by Zwietering et al. (1990) consists in estimating the Gompertz parameters by fitting the following expression to the experimental data:

$$\log[N(t)] = K + A \cdot \exp\left\{-\exp\left\{\left[(\mu_{max} \cdot 2.7182) \cdot \frac{\lambda - t}{A}\right] + 1\right\}\right\}, \quad (1.1)$$

where $N(t)$ is the viable cell concentration (CFU/g) at time t, K is related to the initial level of the viable cell concentration (log CFU/g), A is related to the difference between the decimal logarithm of maximum bacteria growth attained at the stationary phase and the decimal logarithm of the initial value of viable cell concentration, μ_{max} is the maximal specific growth rate, and λ is the lag time. Once the modified Gompertz function parameters are estimated, the SL of the packaged produce is calculated through the following expression:

$$SL = \lambda - \frac{A \cdot \left\{\ln\left[-\ln\left(\frac{Log(N_{max}) - K}{A}\right)\right] - 1\right\}}{\mu_{max} \cdot 2.7182}, \quad (1.2)$$

where N_{max} is the microbial threshold (CFU/g), which, as reported previously, is equal to $5 \cdot 10^7$ CFU/g for minimally processed vegetables. It is worth noting that, even when it is possible to use Eq. 1.1 for estimating the confidence interval of each Gompertz parameter, it is not possible to directly estimate the confidence interval of the SL because it does not appear explicitly as an equation parameter. The difficulty of estimating the SL confidence interval is the main drawback of using the foregoing approach to estimating the SL of fresh produce.

   An alternative method for estimating the SL of fresh-cut products was proposed by Corbo et al. (2006). It consists in rearranging Eq. 1.1 in such a way that the SL parameter appears in the equation, relating the log (CFU/g) to the storage time:

$$\log[N(t)] = \log(N_{max}) - A \cdot \exp\left\{-\exp\left\{\left[(\mu_{max} \cdot 2.71) \cdot \frac{\lambda - SL}{A}\right] + 1\right\}\right\}$$
$$+ A \cdot \exp\left\{-\exp\left\{\left[(\mu_{max} \cdot 2.71) \cdot \frac{\lambda - t}{A}\right] + 1\right\}\right\}. \quad (1.3)$$

   It is worth noting that SL is the time at which the microbiological threshold is reached [i.e., the time at which $N(t)$ is equal to N_{max}].

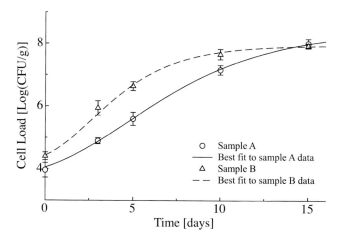

Fig. 1.1 Evolution of mesophilic bacteria as a function of storage time for fresh-cut lettuce. The curves are the best fit of Eq. 1.3 to the experimental data. *Sample A*: treated lettuce with a solution containing 150 ppm of free chlorine; *sample B*: treated lettuce with a solution containing 100 ppm of free chlorine and washed after cutting

By fitting Eq. 1.3 to the experimental data it is possible to estimate the equation's parameters and their confidence interval. Therefore, Eq. 1.3 can be used in place of Eqs. 1.1 and 1.2 to determine both the SL and the confidence interval. The model proposed by Corbo et al. (2006) was used for the mesophilic bacteria cell load of different packaged minimally processed vegetables (fresh-cut lettuce, fennel, and shredded carrots). As an example, Fig. 1.1 shows the evolution during storage of the microbial population in samples of packed fresh-cut lettuce, as reported by Corbo et al. (2006). The curves shown in the figure were obtained by the authors by fitting Eq. 1.3 to the experimental data. Table 1.1 reports some examples of SL values for ready-to-eat produce, obtained according to the approach proposed by Corbo et al. (2006). As pointed out earlier, using Eq. 1.3 it is possible to estimate the SL confidence interval, which in turn makes it possible to establish whether or not there is a significant difference in the SL among the packaged products.

1.3   Multiple Quality Index Models

Whenever the quality of a given packaged food depends on various quality indices, its SL is, by definition, the time at which one of the food quality indices reaches its threshold. Packaged fresh dairy products can serve an example of food whose quality has been reported to depend on more than one single quality index. In fact, the numerous works reported in the literature dealing with fresh dairy product SL have determined that food quality is related to both microbial and sensory quality (Conte et al. 2009a; Del Nobile et al. 2009b; Gammariello et al. 2008b;

Table 1.1 Shelf life values obtained according to mathematical model proposed by Corbo et al. (2006). The 95 % confidence intervals of the calculated shelf life values, calculated on the basis of 200 converging interactions, are shown in square brackets

Product[a]		Shelf life (days)
Shredded carrots	C1	6.42 [4.65, 7.15]
	C2	6.92 [6.34, 7.69]
	C3	6.30 [5.65, 7.00]
	C4	4.56 [4.27, 4.86]
Fresh-cut lettuce	L1	12.63 [11.5, 14.7]
	L2	–[b]
	L3	9.69 [8.47, 11.7]
	L4	4.51 [3.62, 5.41]

[a]C1, C2, C3, C4: shredded carrots produced according to processing lines I, II, III, and IV, respectively; L1, L2, L3, and L4: fresh-cut lettuce produced according to processing lines I, II, III and IV
I: treatment with solution containing 150 ppm of free chlorine
II: treatment with solution containing 100 ppm of free chlorine
III: treatment with a solution containing 100 ppm of free chlorine and washing after cutting for lettuce or shredding for carrots to reduce the residual chlorine concentration
IV: pause of 12 h at room temperature (15–18 °C) before treatment with a chlorine solution (100 ppm of free chlorine) and washing
[b]Mesophilic bacterial count that did not attain $5 \cdot 10^7$ CFU/g

Papaioannou et al. 2007; Pintado et al. 2001). Generally, for assessing dairy microbial quality *Pseudomonas* spp. and coliforms are used as target microbial groups (Conte et al. 2009a; Del Nobile et al. 2009b; Gammariello et al. 2008b). In several cases, the quality of fresh-cut produce has also been described by means of more than one quality index (Mastromatteo et al. 2009; Watada and Qi 1999). Various studies dealing with ready-to-use vegetables (lampascioni, artichokes, and zucchini) also took into account microbial and sensory quality for the purpose of assessing SL (Conte et al. 2009b; Del Nobile et al. 2009a; Lucera et al. 2010).

The way the aforementioned quality indices are described through mathematical models depends on the trend of the experimental data. In fact, it must be recalled that direct models are empirical in nature, and usually their parameters have no particular physical meaning; they are only used to interpolate data by means of a fitting procedure. Therefore, the choice of model should be based solely on its simplicity and its ability to fit the experimental data. Several empirical models can be found in the literature to describe packaged-food quality. For instance, Gammariello et al. (2008a) proposed a first-order kinetic type of equation to quantitatively determine the influence of chitosan on the sensory quality decay of Apulia spreadable cheese during storage. To derive the model, the authors started from a first-order equation:

$$\frac{d\xi(x)}{dx} = -k \cdot x, \qquad (1.4)$$

where ξ is the normalized dependent variable, x is the generic independent variable, and k is the kinetic parameter; ξ is defined as follows:

$$\xi(x) = \frac{y(x) - y^{\infty}}{y^0 - y^{\infty}}, \tag{1.5}$$

where $y(x)$ is the generic dependent variable, y^0 is the initial value of $y(x)$, and y^{∞} is the asymptotic value of $y(x)$. The solution of Eq. 1.4 is as follows:

$$\xi(x) = \xi^0 \cdot \exp(-k \cdot x). \tag{1.6}$$

Substituting Eq. 1.5 into Eq. 1.6 one obtains

$$y(x) = y^{\infty} + (y^0 - y^{\infty}) \cdot \exp(-k \cdot x). \tag{1.7}$$

Equation 1.7 can be further rearranged to incorporate as a parameter the threshold value of the dependent value y:

$$
y(x) = \frac{y^T - y^0 \cdot \exp(-k \cdot x^T)}{1 - \exp(-k \cdot x^T)} \\
+ \left[y^0 - \frac{y^T - y^0 \cdot \exp(-k \cdot x^T)}{1 - \exp(-k \cdot x^T)} \right] \cdot \exp(-k \cdot x), \tag{1.8}
$$

where y^T is the threshold value of $y(x)$, and x^T is the value of the independent variable at which $y(x)$ reaches y^T.

Equation 1.8 was used by the authors to interpolate the sensory data of Apulia spreadable cheese. The authors introduced the concept of sensory acceptability limit, rewriting the preceding equation in the following form:

$$
OSQ(t) = \frac{OSQ_{min} - OSQ_0 \cdot \exp(-k \cdot SAL)}{1 - \exp(-k \cdot SAL)} \\
+ \left(OSQ_0 - \frac{OSQ_{min} - OSQ_0 \cdot \exp(-k \cdot SAL)}{1 - \exp(-k \cdot SAL)} \right) \cdot \exp(-k \cdot t), \tag{1.9}
$$

where $OSQ(t)$ is the packaged-food overall sensory quality at time t, OSQ_0 is the initial value of the packaged-food overall sensory quality, OSQ_{min} is the packaged-food overall sensory quality threshold, SAL is the sensorial acceptability limit [i.e., the time at which $SA(t)$ is equal to SA_{min}]. It is worth noting that in all cases where the food quality depends on several indices, the time at which one of its quality indices reaches the threshold does not necessarily coincide with food SL. Consequently, Gammariello et al. (2008a) used the term *sensory acceptability*

Table 1.2 Appearance, texture, flavor, and overall acceptability of spreadable cheese samples studied by Gammariello et al. (2008a)

Samples	Appearance	Texture	Flavor	Overall acceptability
CTRL	13.58 ± 1.79^a	1.67 ± 0.32^c	12.06 ± 1.48^a	10.56 ± 0.65^a
C12	8.55 ± 0.88^b	3.64 ± 1.25^b	9.81 ± 1.64^b	8.10 ± 0.52^b
C24	>18	10.18 ± 1.29^a	>18	>18
C36	7.33 ± 0.81^b	4.46 ± 1.32^b	7.63 ± 0.8^b	3.71 ± 0.86^c

Data are presented $\pm$ standard deviation

[a-c]Data in each column with different superscript letter are statistically different ($p < 0.05$)

CTRL Apulia spreadable cheese without chitosan, *C12* Apulia spreadable cheese with chitosan – final concentration in working milk 0.012 % (wt/vol), *C24* Apulia spreadable cheese with chitosan – final concentration in working milk 0.024 % (wt/vol), *C36* Apulia spreadable cheese with chitosan – final concentration in the working milk 0.036 % (wt/vol)

limit (SAL) to indicate the time at which the sensory quality reaches its threshold. If this attribute is the sole packaged-food quality index, the *SAL* value would coincide with the product SL. The calculated values of each sensory attribute are reported in Table 1.2. As can be inferred, the *SAL* of the most studied Apulia spreadable cheese samples is mainly affected by texture characteristics. Gammariello et al. (2008a) took into account two quality indices, microbial and sensory ones, to evaluate the SL of dairy products. Considering that the monitored viable cell counts of target spoilage microorganisms were found to be below the threshold for the entire observation period, the authors assessed that the *SAL* values coincided with the SL of the Apulia spreadable cheese.

A Weibull type equation can also be used to interpolate the experimental data and calculate the quality acceptability limit of a packaged food. The Weibull equation was originally used to describe the cumulative density function (Park 1979), which has the following form:

$$y = 1 - \exp\left[-\left(\frac{x}{\beta}\right)^{\alpha}\right], \tag{1.10}$$

where α and β are shape and scale parameters, respectively. Equation 1.10 is an increasing sigmoid function ranging between 0 and 1. To use Eq. 1.10 to interpolate a packaged-food quality index, such as the sensory one, it must be modified to a decreasing function ranging between two arbitrary limits:

$$y = K_1 + K_2 \cdot \exp\left[-\left(\frac{x}{\beta}\right)^{\alpha}\right], \tag{1.11}$$

where K_1 and K_2 are fitting parameters. Equation 1.11 is now a decreasing function ranging between $(K_1 + K_2)$ and K_1. Equation 1.11 can be further rearranged to make the quality index acceptability limit (i.e., the time at which the quality index reaches its threshold value) appear directly as a parameter of the equation relating the

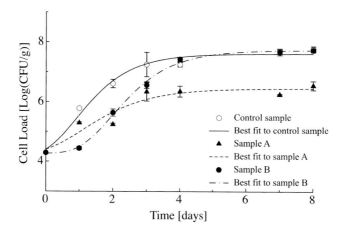

Fig. 1.2 Evolution of *Pseudomonas* spp. as a function of storage time for fior di latte cheese. The curves are the best fit of Eq. 1.13 to the experimental data. *Sample A*: coated cheese sample with sodium alginate (8 %) and 0.25 mg/ml of lysozyme/EDTA; *sample B*: coated cheese sample with sodium alginate (5 %) and 0.25 mg/ml of lysozyme/EDTA

quality index to the storage time. In the specific case where the sensory quality must be interpolated, Eq. 1.11 can be rearranged as follows:

$$OSQ(t) = OSQ_{min} - K_2 \cdot \exp\left[-\left(\frac{SAL}{\beta}\right)^{\alpha}\right] + K_2 \cdot \exp\left[-\left(\frac{t}{\beta}\right)^{\alpha}\right]. \qquad (1.12)$$

A different equation was used by Del Nobile et al. (2010) to quantitatively determine the efficiency of alginate gel loaded with lysozyme and ethylenediaminetetraacetic acid (EDTA) in controlling microbial growth in fior di latte cheese. The authors used the Gompertz equation as reparameterized by Corbo et al. (2006) to interpolate the microbial growth curve:

$$\log[N(t)] = \log(N_{max}) - A \cdot \exp\left\{-\exp\left\{\left[(\mu_{max} \cdot 2.71) \cdot \frac{\lambda - MAL}{A}\right] + 1\right\}\right\}$$
$$+ A \cdot \exp\left\{-\exp\left\{\left[(\mu_{max} \cdot 2.71) \cdot \frac{\lambda - t}{A}\right] + 1\right\}\right\},$$

$$(1.13)$$

where MAL is the microbial acceptability limit, i.e., the time at which the microbiological threshold is reached [i.e., the time at which $N(t)$ is equal to N_{max}]. It is worth noting that the term *microbial acceptability limit* (MAL) has been used in place of *shelf life*, which was used in Eq. 1.3. Figure 1.2 presents an example of the fitting ability of Eq. 1.13. The figure shows the evolution during storage of *Pseudomonas* spp. viable cell concentration in some fior di latte cheese samples along with the best fit of Eq. 1.13 to the experimental data. The authors set

Table 1.3 Shelf life of fior di latte cheese samples evaluated as lowest value between MAL and SAL, calculated as fitting parameters

Sample	Shelf life (days)
Control sample	1.33 ± 0.12^a
0.25 mg ml^{-1} lysozyme and 50 mM Na$_2$–EDTA in brine	3.37 ± 0.14^b
0.50 mg ml^{-1} lysozyme and 50 mM Na$_2$–EDTA in brine	3.16 ± 0.15^b
1.00 mg ml^{-1} lysozyme and 50 mM Na$_2$–EDTA in brine	2.71 ± 0.29^e
0.25 mg ml^{-1} lysozyme and 50 mM Na$_2$–EDTA in alginate (8%) coating	2.74 ± 0.52^e
0.50 mg ml^{-1} lysozyme and 50 mM Na$_2$–EDTA in alginate (8%) coating	2.72 ± 0.00^e
1.00 mg ml^{-1} lysozyme and 50 mM Na$_2$–EDTA in alginate (8%) coating	$2.51 \pm 0.05^{d,e}$
0.25 mg ml^{-1} lysozyme and 50 mM Na$_2$–EDTA in alginate (5%) coating	$2.35 \pm 0.07^{d,e}$
0.50 mg ml^{-1} lysozyme and 50 mM Na$_2$–EDTA in alginate (5%) coating	$2.16 \pm 0.24^{c,d}$
1.00 mg ml^{-1} Lysozyme and 50 mM Na$_2$–EDTA in alginate (8 %) coating	1.91 ± 0.09^c

Data are presented $\pm$ standard deviation
[a-e]Data in column with different superscript letters are significantly different ($p < 0.05$)

the value of N_{max} for *Pseudomonas* spp. at 10^6 CFU/g because at this contamination level alterations of the product start to appear (Bishop and White 1986). The authors used the Gompertz equation as reparameterized by Corbo et al. (2006) also to quantitatively determine the efficiency of the packaging system in slowing down the quality loss of fior di latte cheese in terms of sensory quality preservation:

$$OSQ(t) = OSQ_{min} - A^Q \cdot \exp\left\{-\exp\left\{\left[(\mu_{max}^Q \cdot 2.71) \cdot \frac{\lambda^Q - SAL}{A^Q}\right] + 1\right\}\right\}$$
$$+ A^Q \cdot \exp\left\{-\exp\left\{\left[(\mu_{max}^Q \cdot 2.71) \cdot \frac{\lambda^Q - t}{A^Q}\right] + 1\right\}\right\},$$

$$(1.14)$$

where A^Q is related to the difference between the packaged-food overall sensory quality attained at the stationary phase and the initial value of packaged-food overall sensory quality, μ^Q_{max} is the maximal rate at which $OSQ(t)$ changes, and λ^Q is the lag time. Values of SL were calculated as the lowest value between the MAL and the SAL values (Table 1.3).

A further example of the use of empirical equations to calculate the SL of packaged food is provided by Gammariello et al. (2011), who conducted a study on the effects of the addition of chitosan during cheese making, combined with modified atmosphere packaging (MAP) to prolong the SL of stracciatella cheese stored at 4 °C. The authors calculated the stracciatella cheese SL using three quality parameters, two related to the growth of two spoilage microbial groups, *Pseudomonas* spp. and total coliforms, and the third related to the sensory quality of packaged food. Equations 1.13 and 1.14, used by Gammariello et al. (2011), interpolate the experimental data in a quite acceptable way. The SAL and MAL values calculated by the authors were compared to determine stracciatella cheese SL (Table 1.4).

Table 1.4 Shelf life of stracciatella cheese samples evaluated as lowest value between MAL and SAL, calculated as fitting parameters

Sample	SAL (days)	MAL (days)	Shelf life (days)
Cnt	4.3 ± 0.2^{b}	3.4 ± 0.2^{a}	3.4 ± 0.2^{a}
Ch1	$5.2 \pm 0.3^{a,b}$	4.0 ± 0.2^{a}	4.0 ± 0.2^{a}
Ch2	$4.9 \pm 0.2^{a,b}$	3.9 ± 0.3^{a}	3.9 ± 0.3^{a}
Ch3	$6.0 \pm 0.4^{a,c}$	3.7 ± 0.2^{a}	3.7 ± 0.2^{a}
Cnt-MAP	7.5 ± 0.3^{d}	5.8 ± 0.3^{b}	5.8 ± 0.3^{b}
Ch1-MAP	5.8 ± 0.3^{a}	6.5 ± 0.2^{c}	5.8 ± 0.3^{b}
Ch2-MAP	$6.9 \pm 1.1^{c,d}$	7.0 ± 0.6^{c}	6.9 ± 1.1^{c}
Ch3-MAP	5.6 ± 1.0^{a}	6.8 ± 0.5^{c}	5.6 ± 1.0^{b}

Data are presented $\pm$ standard deviation

[a-d]Data in column with different small letters are significantly different ($p < 0.05$)

Cnt chitosan-free stracciatella packaged in tubs, *Ch1* stracciatella with chitosan added to 0.010 % and packaged in tubs, *Ch2* stracciatella with chitosan added to 0.015 % and packaged in tubs, *Ch3* stracciatella with chitosan added to 0.020 % and packaged in tubs, *Cnt MAP* chitosan-free sample packaged under MAP, *Ch1-MAP* stracciatella with chitosan added to 0.010 % and packaged under MAP, *Ch2-MAP* stracciatella with chitosan added to 0.015 % and packaged under MAP, *Ch3-MAP* stracciatella with chitosan added to 0.020 % and packaged under MAP

MAL microbiological acceptability limit

SAL sensorial acceptability limit

References

Baranyi J, Roberts TA (1995) Mathematics of predictive food microbiology. Int J Food Microbiol 26:199–218

Bishop JR, White CH (1986) Assessment of dairy product quality and potential shelf life-a review. J Food Prot 49:739–753

Conte A, Gammariello D, Di Giulio S, Attanasio M, Del Nobile MA (2009a) Active coating and modified atmosphere packaging to extend the shelf life of fior di latte cheese. J Dairy Sci 92:887–894

Conte A, Scrocco C, Brescia I, Del Nobile MA (2009b) Packaging strategies to prolong the shelf life of minimally processed lampascioni (*Muscari comosum*). J Food Eng 90:199–206

Corbo MR, Altieri C, D'Amato D, Campaniello D, Del Nobile MA, Sinigaglia M (2004) Effect of temperature on shelf-life and microbial population of lightly processed cactus pears fruit. Postharvest Biol Technol 31:93–104

Corbo MR, Del Nobile MA, Sinigaglia M (2006) A novel approach for calculating shelf life of minimally processed vegetables. Int J Food Microbiol 106:69–73

Del Nobile MA, Conte A, Scrocco C, Laverse J, Brescia I, Conversa G, Elia A (2009a) New packaging strategies to preserve fresh-cut artichoke quality during refrigerated storage. Innov Food Sci Emerg Technol 10:128–133

Del Nobile MA, Gammariello D, Conte A, Attanasio M (2009b) A combination of chitosan, coating and modified atmosphere packaging for fior di latte cheese. Carbohydr Polym 78:151–156

Del Nobile MA, Gammariello D, Di Giulio S, Conte A (2010) Active coating to prolong the shelf life of fior di latte cheese. J Dairy Res 77:144–150

Gammariello D, Chillo S, Mastromatteo M, Di Giulio S, Attanasio M, Del Nobile MA (2008a) Effect of chitosan on the rheological and sensorial characteristics of Apulia spreadable cheese. J Dairy Sci 91:4155–4163

Gammariello D, Di Giulio S, Conte A, Del Nobile MA (2008b) Effects of natural compounds on microbial safety and sensory quality of fior di latte cheese, a typical Italian cheese. J Dairy Sci 91:4138–4146

Gammariello D, Conte A, Attanasio M, Del Nobile MA (2011) A study on the synergy of modified atmosphere packaging and chitosan on stracciatella shelf life. J Food Process Eng 34:1394–1407

Lanciotti R, Corbo MR, Gardini F, Sinigaglia M, Guerzoni ME (1999) Effect of hexanal on the shelf-life of fresh apple slices. J Agr Food Chem 47:4769–4776

Lucera A, Costa C, Mastromatteo M, Conte A, Del Nobile MA (2010) Influence of different packaging systems on fresh-cut zucchini (Cucurbita pepo). Innov Food Sci Emerg Technol 11:361–368

Mastromatteo M, Conte A, Del Nobile MA (2009) Preservation of fresh-cut produce using natural compounds. Review article. Stewart Postharvest Rev 4:4

Ministere de l'Economie des Finances et du Budget (1988) Marché consommation, produits vegetaux prêts à l'emploi dits de la "IVemme gamme": guide de bonnes pratique hygiéniques. Journal Officiel de la République Française 1621:1–29

Papaioannou G, Chouliara I, Karatapanis AE, Kontominas MG, Savvaidis IN (2007) Shelf-life of a Greek whey cheese under modified atmosphere packaging. Int Dairy J 17:358–364

Park WJ (1979) Basic concepts of statistics and their applications in composite materials. AFML-TR-79-4070. Wright-Patterson AFB, Ohio, pp 25–28

Pintado ME, Macedo AC, Malcata FX (2001) Review: technology, chemistry and microbiology of whey cheeses. Food Sci Technol Int 7:105–116

Riva M, Franzetti L, Galli A (2001) Microbiological quality and shelf life modeling of ready-to-eat cicorino. J Food Prot 64(2):228–234

Sinigaglia M, Corbo MR, D'Amato D, Campaniello D, Altieri C (2003) Shelf-life modelling of ready-to-eat coconut. Int J Food Sci Technol 38:547–552

Watada A, Qi L (1999) Quality of fresh-cut produce. Postharvest Biol Technol 15:201–205

Zwietering MH, Jongenburger FM, Roumbouts M, van't Riet K (1990) Modelling of the bacterial growth curve. Appl Environ Microbiol 56:1875–1881

Zwietering MH, De Koos JT, Hasenack BE, De Wit JC, van't Riet K (1991) Modelling of bacterial growth as a function of temperature. Appl Environ Microbiol 57:1875–1881

Chapter 2
Influence of Mass Transport Properties of Films on the Shelf Life of Packaged Food

2.1 Introduction

The barrier properties of flexible films play a major role in determining the shelf life (SL) of packed foodstuffs (Quast and Karel 1972; Talasila and Cameron 1997). In fact, polymeric films controlling the rate at which small molecular weight compounds permeate into or outside a package can slow down many of the detrimental phenomena responsible for the packaged product's unacceptability. Generally, the mass transport properties of polymeric films are simply determined by evaluating the permeability coefficient of a given polymeric-diffusant system (Del Nobile et al. 1996a, b). This approach can be successfully used wherever the permeability coefficient does not depend on the packaging film boundary conditions. For instance, this is the case with gas permeation through rubbery films such as polyethylene. In fact, in these cases, the coefficient is constant because the permeating molecules do not change, to a great extent, the free volume of the polymeric matrix.

The foregoing approach fails wherever the permeability depends on the diffusant partial pressure at the upstream and downstream sides of the film, as in the case of water vapor transmission through moderately hydrophilic polymers. In fact, water molecules acting as plasticizers increase the macromolecular mobility of the polymer. As a consequence, both the solubility and diffusivity coefficients, and consequently the permeability coefficient, depend on water activity. In these cases, the permeability coefficient cannot be determined by a single measure; in its place, a more accurate analysis of the permeation process is necessary to properly determine the barrier properties of the film. In the following paragraphs, mathematical models to predict the barrier properties of single multilayer systems as a function of the water activity at the upstream and downstream sides of a film will be presented and discussed.

M.A. Del Nobile and A. Conte, *Packaging for Food Preservation*,
Food Engineering Series, DOI 10.1007/978-1-4614-7684-9_2,
© Springer Science+Business Media New York 2013

2.2 Single Layer Structures

Several works are reported in the literature dealing with the influence of water on the permeability of gases through water-sensitive packaging film. Gavara and Hernandez (1994) described oxygen permeability experiments by a bimodal diffusion mechanism. The two processes were called fast and slow, and the total permeability was expressed as a linear combination of these two mechanisms; from the values of permeability and diffusion, the solubility of oxygen in nylon-6 was calculated for both mechanisms. The change in the oxygen permeability as a function of polymer moisture content was controlled by the change in oxygen solubility as a function of moisture, which was related to the state of water in the polymer matrix. The authors assumed that this dependency was caused by the formation of a water cluster and the consequent molecular competition between water and oxygen for the polymer active sites. Del Nobile et al. (2003b) proposed a mechanistic approach to describing the water barrier properties of nylon as a moderately hydrophilic polymer in which both solubilization and a diffusion process were separately described. The proposed mathematical model was validated by successfully predicting the water permeability coefficient of a nylon film. However, models that describe separately the main phenomena involved in the mass transport process are somewhat complex to handle from a calculative point of view. Moreover, they need a rather complex and expensive experimental procedure to obtain the values of the model parameters. An alternative to this approach is represented by empirical mathematical models that are generally easier to handle and the parameters are simply determined. These mathematical systems can be applied to several moderately hydrophilic polymeric films, and in some circumstances they can also be advantageously used for packaging-design purposes.

In what follows, both mechanistic and empirical approaches to predicting the barrier properties of single-layer structures as a function of water activity at the upstream and downstream sides of a film are presented and discussed separately.

2.2.1 Empirical Models

Regarding the dependence of low molecular weight compound permeability on relative humidity at film boundaries, it either increases or decreases, or it shows an upward concavity. A steady increase in the permeability coefficient with relative humidity might be found when water molecules acting as a plasticizer increase the macromolecular mobility of a rubbery polymer. On the other hand, the permeability coefficient might steadily decrease with relative humidity whenever water molecules competing with the permeating compound reduce the frozen microvoid capacity of a glassy polymer to host the permeating substance (Del Nobile et al. 1997a, 2003a). An upward concavity of the permeability coefficient plotted as a function of relative humidity at the boundaries can be encountered in the case of glassy polymers. In these cases, water molecules can reduce the low molecular weight compound solubility coefficient by reducing the frozen microvoid capacity to host the permeating compound; on the other hand, they can act as a plasticizer and increase the macromolecular mobility, thereby increasing the low molecular weight compound diffusivity. In fact, an upward concavity means that the former effect prevails at a lower relative humidity, whereas the latter one prevails at higher values.

An example of an empirical model is presented by Mastromatteo and Del Nobile (2011), who proposed a simple approach to predicting the oxygen permeability coefficient of two films made up of polyethylenterephthalate (PET) and oriented polyamide (OPA) as a function of the relative humidity at the boundaries. The authors proposed the following two empirical equations to account for all the aforementioned possible dependence of the permeability coefficient on relative humidity:

$$P_{O_2}\left(\overline{\%RH}\right) = P_{O_2}^{\infty} + \left(P_{O_2}^{0} - P_{O_2}^{\infty}\right) \cdot \exp\left(-b_0 \cdot \overline{\%RH}\right), \qquad (2.1)$$

$$P_{O_2}\left(\overline{\%RH}\right) = P_{O_2}^{0} + b_1 \cdot \overline{\%RH} + b_2 \cdot \overline{\%RH}^2, \qquad (2.2)$$

where $\overline{\%RH}$ is the average relative humidity across the film, $P_{O_2}^{0}$ is the film oxygen permeability at zero $\overline{\%RH}$, $P_{O_2}^{\infty}$ is the hypothetical film oxygen permeability reached at an infinite value of $\overline{\%RH}$, and b_i are constants that account for the dependence of oxygen permeability on the average relative humidity across the film. In fact, Eq. 2.1 is a first-order kinetic-type equation and can be used whenever the oxygen permeability steadily either increases or decreases with $\overline{\%RH}$. On the other hand, Eq. 2.2 is a second-order polynomial function to be used whenever an upward concavity is found.

As reported earlier, Mastromatteo and Del Nobile (2011) used PET and OPA films to validate Eqs. 2.1 and 2.2. Figure 2.1 shows the oxygen permeability coefficient plotted as a function of the average relative humidity across the film for PET and OPA, respectively. The curve shown in each picture of Fig. 2.1 was obtained by fitting Eqs. 2.1 and 2.2 to the experimental data. The model parameters calculated from the fitting are as follows: $P_{O_2}^{\infty} = 1.90 \cdot 10^{-11} \pm 1.46 \cdot 10^{-12}$; $P_{O_2}^{0} = 2.97 \cdot 10^{-11} \pm 4.87 \cdot 10^{-13}$; $b_0 = 1.93 \cdot 10^{-2} \pm 5.44 \cdot 10^{-3}$ for the PET film; $P_{O_2}^{0} = 1.82 \cdot 10^{-11} \pm 2.15 \cdot 10^{-12}$; $b_1 = -6.75 \cdot 10^{-13} \pm 9.39 \cdot 10^{-14}$; $b_2 = 9.08 \cdot 10^{-15} \pm 9.91 \cdot 10^{-16}$ for the OPA film.

As can be inferred from the figure, the empirical expression proposed by Mastromatteo and Del Nobile (2011) interpolate the experimental data in a quite acceptable way; however, other insights into the mechanism of low molecular weight compound permeation through the polymeric film are possible.

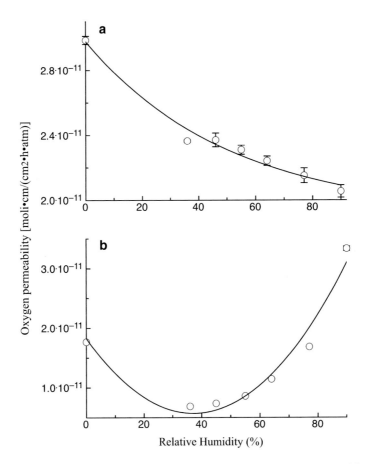

Fig. 2.1 Oxygen permeability coefficient plotted as function of average relative humidity across film for PET (**a**) and OPA (**b**). The *curve* shown in each picture was obtained by fitting Eqs. 2.1 and 2.2, respectively, to the experimental data

2.2.2 Mechanistic Models

A mechanistic approach to this problem needs to start from the phenomena involved in the permeation of low molecular weight compounds through polymeric films. It is generally accepted that this mechanism consists of three steps: (1) the absorption of the low molecular weight compound at the high partial pressure side of the film, (2) its diffusion through the polymeric matrix, and (3) its desorption at the low pressure side of the film. Therefore, the permeation through a polymeric matrix is a combination of two distinct processes: solubilization and diffusion.

Normally, low molecular weight compound diffusion through a film is described by Fick's model, which for a plane sheet has the following form:

$$J = -D(C) \cdot \frac{\partial C}{\partial x},\qquad(2.3)$$

where J is the low molecular weight compound mass flux, D(C) is the diffusion coefficient, C is the low molecular weight compound local concentration, and x is the spatial coordinate. Integrating Eq. 2.3 across the film thickness one obtains

$$J_{SS} \cdot 1 = -\int_0^1 \left(D(C) \cdot \frac{\partial C}{\partial x}\right) \cdot dx = \int_{C_2}^{C_1} D(C) \cdot dC,\qquad(2.4)$$

where C_1 and C_2 are the permeant concentration at the upstream and downstream sides of the film, respectively. Equation 2.4 was derived by setting the origin of the spatial coordinate at the high partial pressure side of the film (i.e., upstream side of the film) and assuming that steady-state conditions are obeyed. Comparing Eq. 2.4 with Eq. 2.3 one obtains

$$P(p_1, p_2) = \frac{1}{p_1 - p_2} \cdot \int_{C_2(p_2)}^{C_1(p_1)} D(C) \cdot dC,\qquad(2.5)$$

where p_1 and p_2 are the permeant partial pressure at the upstream and downstream sides of the film, respectively. It is worth noting that Eq. 2.5 provides the relationship between the permeability coefficient and the permeant partial pressure at the film sides, and it is derived directly from the assumption of permeation as a three-step process.

Water vapor is one of the most important low molecular weight compounds of a specific group of flexible films intended for food-packaging applications that can make the permeability coefficient dependent on its partial pressure at the film boundaries. Therefore, in what follows, the focus of this paragraph will be on the dependence of the permeability coefficient on the water vapor partial pressure at the film sides. In the specific case of water permeation, Eq. 2.5 can be rewritten as follows:

$$P_W\left(p_W^1, p_W^2\right) = \frac{1}{p_W^1 - p_W^2} \cdot \int_{C_W\left(p_W^2\right)}^{C_W\left(p_W^1\right)} D_W(C_W) \cdot dC_W,\qquad(2.6)$$

where P_W is the water vapor permeability coefficient, p_W^1 and p_W^2 are the water vapor partial pressure at the upstream and downstream sides of the film, respectively, C_W is the water concentration, and D_W is the water vapor diffusion coefficient.

To use Eq. 2.6, the following information must be obtained: (1) the relationship between the water concentration in the polymer at equilibrium and its partial pressure, which is a quantitative description of the solubilization process, and (2) the dependence of water diffusivity on the local water concentration, which is a quantitative description of the diffusion process. These two distinct aspects of water permeation through moderately hydrophilic polymers will be described for two distinct cases: (1) where the polymer matrix swelling can be neglected and (2) where the polymeric swelling and polymeric matrix relaxation must be taken into account. Finally, a third case study will be presented in which the relationship between the oxygen permeability and the water vapor partial pressure at the boundary is also considered.

2.2.2.1 Water Permeability: Negligible Polymeric Swelling

Water sorption in moderately hydrophilic polymers is a rather complex phenomenon due to specific interactions between water molecules and hydrophilic sites on the polymer backbone. Consequently, the absorbed water molecules are in part randomly dispersed into the polymer matrix (dissolved water or free water) and in part physically bonded to the hydrophilic sites (adsorbed water or bound water) (Netti et al. 1996). As first approximation it can be assumed that the amount of adsorbed water is negligible if compared to the amount of dissolved water. As long as the preceding hypothesis is valid, the equation proposed by Flory (1953) to describe the mixing of a linear polymer with a low molecular weight compound can be used to relate water concentration to water activity:

$$\ln(a_W) = \ln\left(\frac{C_W}{\rho_W + C_W}\right) + \left(1 - \frac{C_W}{\rho_W + C_W}\right) + \chi \cdot \left(1 - \frac{C_W}{\rho_W + C_W}\right)^2, \quad (2.7)$$

where a_W is the water activity calculated as the ratio between the water vapor partial pressure and the equilibrium water vapor, ρ_W is the water density, and χ is the Flory interaction parameter that measures the affinity between the polymer and the dissolved low molecular weight compound. Generally, ρ_W is either directly measured or taken from the literature, whereas χ is obtained by fitting Eq. 2.7 to the sorption isotherm data. The fitting process has two main objectives: to prove that the equation is able to interpolate the experimental data and to determine the model parameter (χ).

Del Nobile et al. (2002) used the approach illustrated earlier to describe the solubilization of water in cellophane film. As an example, Fig. 2.2 shows the equilibrium water concentration in cellophane plotted as a function of water activity. In the same figure the best fit of Eq. 2.7 to the experimental data is also reported. The value of the calculated χ parameter was 0.654. As is evident, the model proposed by Flory (Eq. 2.7) satisfactorily fits the data, suggesting that in

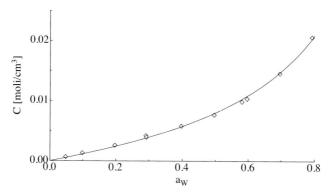

Fig. 2.2 Equilibrium water concentration plotted as function of water activity. The *curve* in the figure represents the best fit of Eq. 2.7 to the experimental data

certain cases, as in this one of a water/cellophane system, the amount of adsorbed water can be neglected if compared to the amount of dissolved water.

Water diffusion through moderately hydrophilic polymers is generally described using Fick's model having a water-concentration-dependent diffusion coefficient. For water diffusion through a plane sheet, Fick's model can be converted into this expression:

$$J = -D_W(C_W) \cdot \frac{\partial C_W}{\partial x}. \tag{2.8}$$

Many equations can be found in the literature to describe the dependence of a water diffusion coefficient on the local water concentration, and most of them are empirical in nature. In what follows, the relationship proposed by Del Nobile et al. (2002) to describe water diffusion in cellophane is reported:

$$D_W(C_W) = A_1 - A_2 \cdot \exp\left[-\left(\frac{C_W}{A_3}\right)^{A_4}\right], \tag{2.9}$$

where A_i are constants without any specific physical meaning and are to be regarded as fitting parameters. Analogously to what was done to determine the χ value (Eq. 2.7), the A_i values are determined by data fitting. Wherever the polymer swelling can be neglected, the approach proposed by Del Nobile et al. (2002) to calculated the A_i values is one of the most widely used because it combines simplicity and accuracy of results. It consists in measuring the water sorption kinetics by increasing stepwise the water vapor partial pressure. The difference between the initial and final water vapor partial pressures must be sufficiently small to ensure the constancy of the water diffusion coefficient during each water sorption test. Afterward, the water sorption kinetic data are fitted using the following expression:

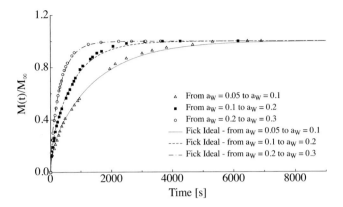

Fig. 2.3 Sorption kinetic curves of water in cellophane film plotted as function of time. The *curves* are the best fit of Eq. 2.10 to the water sorption kinetic data

$$\frac{M(t)}{M_\infty} = 1 - \sum_{n=0}^{\infty} \left\{ \frac{8}{(2 \cdot n + 1)^2 \cdot \pi^2} \cdot \exp\left[-\frac{D_W \cdot (2 \cdot n + 1)^2 \cdot \pi^2 \cdot t}{4 \cdot l^2} \right] \right\}, \quad (2.10)$$

where $M(t)$ is the amount of water absorbed into the polymer at time t, M_∞ is the amount of water absorbed into the polymer at equilibrium, and l is the film thickness. Equation 2.10 is the analytical solution of Fick's second law solved under the following conditions: (1) plane sheet geometry, (2) uniform initial conditions, (3) same water concentration at the plane sheet sides, and (4) constant boundary conditions. As an example, Fig. 2.3 shows some of the sorption kinetic curves of water in cellophane film as measured by Del Nobile et al. (2002), where $\frac{M(t)}{M_\infty}$ is plotted as a function of time. In the same figure the best fit of Eq. 2.10 to the water sorption kinetic data is also presented. Equation 2.10 demonstrates the ability of the model to interpolate the data and, on the other hand, represents a way to calculate the water diffusion coefficient, which is the sole fitting parameter.

Once the water diffusion coefficient is calculated for each of the water sorption kinetics, its value is associated to the average water concentration of the corresponding water sorption kinetic test, which in turn is defined as the average between the initial and final water concentration in each water sorption kinetic test.

As an example, Fig. 2.4 shows the water diffusion coefficient, calculated by the same Eq. 2.10, as a function of average water concentration. In this figure the best fit of Eq. 2.9 to the data is also presented. As can be observed, the interpolative ability of Eq. 2.9 is fairly acceptable. The calculated constants are as follows: $A_1 = 4.12 \cdot 10^{-9} \frac{cm^2}{s}$, $A_2 = 3.73 \cdot 10^{-9} \frac{cm^2}{s}$, $A_3 = 4.31 \cdot 10^{-3} \frac{moli}{cm^3}$, $A_4 = 2.44$.

As long as the parameters appearing in Eqs. 2.7 and 2.8 are known, the water permeability coefficient can be predicted by means of Eq. 2.6. Del Nobile et al. (2002) obtained a good agreement between measured and predicted data for the water permeability of cellophane film, thus assessing the predictive efficacy of the proposed mechanistic model.

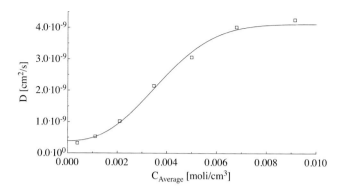

Fig. 2.4 Water diffusion coefficient (D) plotted as function of average water concentration. The *curve* is the best fit of Eq. 2.9 to the experimental data

2.2.2.2 Water Permeability: Polymer Matrix Relaxation

As reported earlier, due to the presence of specific interactions between water molecules and hydrophilic sites on the polymer backbone of moderately hydrophilic polymers, the absorbed molecules are partially dissolved and partially adsorbed (Netti et al. 1996). Wherever the adsorbed water cannot be neglected, the total amount of absorbed water must be expressed as follows:

$$C_W = C_W^D + C_W^{ad}, \qquad (2.11)$$

where C_W is the total amount of absorbed water, C_W^D is the amount of dissolved water, and C_W^{ad} is the amount of adsorbed water. As noted previously, the equation proposed by Flory (1953) to describe the mixing process of a linear polymer with a low molecular weight compound can be successfully used to relate C_W^D to water activity:

$$\ln(a_W) = \ln\left(\frac{C_W^D}{\rho_W + C_W^D}\right) + \left(1 - \frac{C_W^D}{\rho_W + C_W^D}\right) + \chi \cdot \left(1 - \frac{C_W^D}{\rho_W + C_W^D}\right)^2. \qquad (2.12)$$

The literature reports the Langmuir equation as a simple way to describe the dependence of C_W^{ad} on water activity (Netti et al. 1996):

$$C_W^{ad} = \frac{C_H' \cdot b \cdot a_W}{1 + b \cdot a_W}, \qquad (2.13)$$

where C_H' is the hydrophilic site adsorption capacity, and b is a measure of the affinity between water molecules and hydrophilic sites.

Combining Eqs. 2.7 and 2.8 it is possible to relate C_W to water activity. However, due to the structure of Eq. 2.7, it is not possible to combine the preceding equations to obtain a relationship where C_W is an explicit function of a_W. In fact, for a given a_W the corresponding value of C_W must be evaluated by first numerically solving Eq. 2.7 to determine C_W^D, then evaluating C_W^{ad} through Eq. 2.8, and finally summing these two terms.

Del Nobile et al. (2003b) used the approach illustrated earlier to describe the solubilization process of water in nylon film. The values of parameters appearing in Eqs. 2.12 and 2.13, obtained by fitting Eq. 2.11 to the experimental data, are C_H' = $3.23 \cdot 10^{-3}$ $\frac{\text{g Water}}{\text{cm}^3 \text{ Dry Polymer}}$; b = 47.5; $\chi = 1.81$. Del Nobile et al. (2003b) evaluated the goodness of fit by means of the relative percentage difference (Boquet et al. 1978), which is defined by the following expression:

$$\bar{E}\% = \frac{100}{N} \cdot \sum_{i=1}^{N} \frac{|M_i - M_p|}{M_i}, \tag{2.14}$$

where $\bar{E}\%$ is the relative percentage difference, N is the number of experimental points, and M_i and M_p are the experimental and predicted values, respectively. The authors obtained a value of $\bar{E}\%$ equal to 3.22. The good agreement between predicted and experimental data allowed the authors to conclude that the sorption process of water into a nylon film can be described in terms of the hypothesized dual sorption mechanism. Although it is not the easiest way to describe the solubilization process of water in moderately hydrophilic polymers, it is a good compromise between simplicity and rigor of physical frame to derive it.

The diffusion of low molecular weight compounds in polymers is generally governed by two simultaneously occurring phenomena. One is a substantially stochastic process (related to Brownian motion), where the compound flows exclusively driven by a concentration gradient. Another one is the relaxation phenomenon, driven by the distance of the local system to equilibrium (Del Nobile et al. 1994). When a mass transfer takes place in a substantially unperturbed matrix, as in the case of gas diffusion in rubbery polymers (or wherever the solvent-induced polymer swelling is negligible), the diffusion process is largely controlled by a stochastic phenomenon. The other limiting behavior is encountered when a very thin slab of polymer is put in contact with a swelling low molecular weight compound. In this case the characteristic diffusion time is much lower than the polymer relaxation time; hence, the relaxation becomes the limiting phenomenon controlling the solvent uptake kinetic (Del Nobile et al. 1994). In the case of water diffusion in moderately hydrophilic polymers, the experimental observations range between these two limiting phenomena. To better illustrate these two distinct aspects, stochastic diffusion and polymer relaxation will be presented separately.

Water diffusion related to Brownian motions is generally described by means of Fick's model with a diffusion coefficient depending on the local water concentration (Eq. 2.8). As reported by Del Nobile et al. (1997b), a simple expression to relate DW to the local penetrant concentration can be obtained by rearranging the relationship proposed by Fujita (1961) to describe the dependence of the thermo-dynamic diffusion coefficient on the local penetrant volume fraction:

$$D_W(C_W) = B_1 \cdot \exp\left(\frac{-1}{B_2 + B_3 \cdot C_W}\right), \tag{2.15}$$

where the B_i are constants without any specific physical meaning and regarded as fitting parameters.

The literature contains several approaches to describing solvent-induced poly-mer relaxation (Del Nobile et al. 1994). Among them, one of the simplest is that proposed by Long and Richman (1960). They assumed that when the external water activity is suddenly changed, the solvent concentration at the polymer surfaces does not instantaneously reach the equilibrium value (as it would if only stochastic diffusion were considered), but it first rapidly increases to a value lower than the equilibrium one, and at a later stage it continues to increase, gradually reaching equilibrium. The instantaneous response of the system to the increase in the outside water activity represents the elastic response of the polymer matrix to an external perturbation. The value of the initial water concentration at the film boundaries, $C_W^B(0)$, depends on both the initial macromolecular mobility and the extent of perturbation. In turn, the former depends on the initial water concentration in the film, $C_W(0)$. Otherwise, the extent of perturbation is related to the difference between the final $C_W(\infty)$ value and the initial water concentration at the film boundary, $[C_W(\infty) - C_W(0)]$. Due to the complex phenomena involved, the fol-lowing empirical relationship was proposed by Del Nobile et al. (2003b) to relate $C_W^B(0)$ to $C_W(0)$ and $[C_W(\infty) - C_W(0)]$:

$$K = \frac{1 - [C_W(0)]^{K_1} \cdot \exp[-K_2 \cdot C_W(0)]}{\exp\left\{K_3 \cdot [C_W(0) - C_W(\infty)]^2\right\}}, \tag{2.16}$$

where K is the normalized initial water concentration at the boundaries of the film defined by the equation $K = \frac{C_W^B(0) - C_W(0)}{C_W(\infty) - C_W(0)}$, which ranges from zero to one; the K_i are constants to be evaluated by fitting the model to the experimental data.

The rate at which the water concentration at the boundaries gradually increases is directly related to the polymeric matrix relaxation kinetic. In the work of Del Nobile et al. (2003b) the following empirical expression is proposed to describe the boundary condition relaxation rate:

$$\frac{d\alpha(t)}{dt} = \left\{\alpha_1 \cdot \exp\left[\alpha_2 \cdot C_W^B(t)\right] \cdot \sqrt{\alpha(t)}\right\} \cdot \{1 - \exp[-(1 - \alpha(t))]\}, \tag{2.17}$$

where $\alpha(t)$ is the normalized water concentration at the boundaries of the film at time t defined through the equation $\alpha(t) = \frac{C_W^B(t) - C_W^B(0)}{C_W(\infty) - C_W^B(0)}$, which ranges from zero to one;

the α_i are constants to be evaluated by fitting the model to the experimental data. It is worth noting that $\alpha(t)$ represents the driving force of the polymer relaxation. As shown by Eq. 2.17, $\frac{d\alpha(t)}{dt}$ is expressed as the product of two terms. The first one, which prevails at the early stage of the sorption process, is the kinetic constant of the polymer relaxation and increases as the polymer macromolecular mobility increases. The second term, which prevails at the later stage of the sorption process, is a measure of the extent of the driving force of the polymer relaxation (i.e., the distance of the system from equilibrium). Therefore, it is a decreasing function of $\alpha(t)$.

The evolution during hydration of the water concentration profile (alternatively, the water uptake kinetic) can be calculated by solving the water mass balance equation. The differential equation reported in the following expression was derived using Eqs. 2.8 and 2.15 to describe the stochastic diffusion and Eqs. 2.16 and 2.17 to describe the polymer matrix relaxation kinetic:

$$\frac{\partial C_W(t)}{\partial t} = \frac{\partial}{\partial x}\left\{\left[B_1 \cdot \exp\left(\frac{-1}{B_2 + B_3 \cdot C_W(t)}\right)\right] \cdot \frac{\partial C_W(t)}{\partial x}\right\}. \qquad (2.18)$$

Equation 2.18 needs to be solved numerically with the following initial and boundary conditions:

$$\begin{cases} C_W = C_W(0) \Rightarrow t = 0; 0 < x < 1, \\ C_W = C_W^B(t) \Rightarrow \forall t; x = 0, x = 1. \end{cases} \qquad (2.19)$$

The model illustrated earlier to describe water diffusion in the moderately hydrophilic films consists of Eqs. 2.16, 2.17, 2.18, and 2.19 and is referred to as the anomalous diffusion model (Del Nobile et al. 2003b). For the sake of completeness, the same authors named the model reported earlier to describe water diffusion in cellophane films as the nonideal Fickian model.

As an example, Fig. 2.5 shows the water uptake kinetics of the nylon film at different water activity values as measured by Del Nobile et al. (2003b). In the same figure the curves representing the best fit of the anomalous diffusion model to the experimental data are also shown. The authors simultaneously fitted the anomalous diffusion model to the available sorption kinetic to prove the model's ability to describe the experimental data and to evaluate the model parameters. It must be emphasized that the two approaches proposed by Del Nobile et al. (2002, 2003b) not only differ in the way the model is derived (only the former one accounts for polymer relaxation), they also differ in the way the parameters used to relate the water diffusion coefficient to local water concentration are calculated. The model parameters obtained to describe water diffusion in nylon are listed in Table 2.1 (the authors calculated $\bar{E}\%$ values for each sorption kinetic). As can be inferred from the data shown in Fig. 2.5 and listed in Table 2.1, to properly describe the water sorption kinetic in moderately hydrophilic polymers, the superposition of

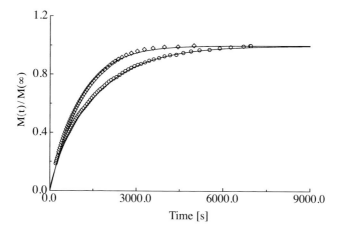

Fig. 2.5 Water uptake kinetics of nylon film at water activity from 0.7 to 0.8 (○) and from 0.8 to 0.9 (◊). The *curves* are the best fit of the anomalous diffusion model to the experimental data

Table 2.1 Values of model parameters obtained by fitting the anomalous diffusion model to the experimental data to describe water diffusion in nylon film

Parameter	Anomalous diffusion model
K_1	2.22
$K_2 \left[\frac{\text{g Dry Polymer}}{\text{g Water}}\right]$	$2.67 \cdot 10^{-2}$
$K_3 \left[\frac{\text{g Dry Polymer}}{\text{g Water}}\right]$	62.8
$\alpha_1 \left[\frac{1}{s}\right]$	$3.87 \cdot 10^{-3}$
$\alpha_2 \left[\frac{\text{g Dry Polymer}}{\text{g Water}}\right]$	9.45
$B_1 \left[\frac{\text{cm}^2}{s}\right]$	$1.37 \cdot 10^{-4}$
B_2	$8.03 \cdot 10^{-2}$
$B_3 \left[\frac{\text{g Dry Polymer}}{\text{g Water}}\right]$	0.196

polymer relaxation to stochastic diffusion must be taken into account, especially at higher water activity. It is also worth noting that by fitting the anomalous diffusion model to the experimental data it is possible to quantitatively resolve each sorption kinetic curve into the two phenomena controlling the sorption process (i.e., stochastic diffusion and polymer matrix relaxation). In this way the parameters appearing in the relationship between the water diffusion coefficient and the local water concentration can be determined with a higher accuracy.

Similarly to cellophane film, once the parameters appearing in Eqs. 2.11 and 2.15 are evaluated, Eq. 2.6 can be used to predict the water permeability coefficient. Figure 2.6 shows the water permeability of nylon film, as determined by means of permeation tests, with the curve obtained using Eq. 2.6 and data listed in Table 2.1. As can be inferred, the ability of the anomalous diffusion model to predict the experimental data is quite acceptable ($\bar{E}\%$ equal to 5.43).

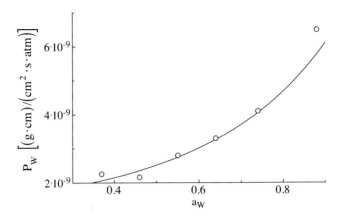

Fig. 2.6 Water permeability of nylon film, as predicted by anomalous diffusion model

2.2.2.3 Oxygen Permeability

In what follows, a mathematical model to predict the oxygen barrier properties of moderately hydrophilic films as a function of the water activity at the upstream and downstream sides of the film is presented. As will be discussed subsequently, to determine the oxygen permeability coefficient, first the steady-state water concentration profile must be evaluated. To this end, both water solubilization and diffusion need to be quantitatively described. According to what was reported earlier, the oxygen permeability coefficient and the steady-state oxygen mass flux are related through the following equation:

$$J_{O_2} = P_{O_2}\left(p_W^1, \; p_W^2\right) \cdot \frac{\Delta p_{O_2}}{l}, \qquad (2.20)$$

where J_{O_2} is the steady-state oxygen mass flux, $P_{O_2}\left(p_W^1, \; p_W^2\right)$ is the oxygen permeability coefficient, and $\frac{\Delta p_{O_2}}{l}$ is the oxygen partial pressure gradient across the film. The oxygen mass flux can be related to the oxygen concentration gradient assuming that the oxygen diffusion process can be described by means of Fick's model, with an oxygen diffusion coefficient depending only on the local water concentration:

$$J_{O_2} = -D_{O_2}(C_W) \cdot \frac{dC_{O_2}}{dx}, \qquad (2.21)$$

where $D_{O_2}(C_W)$ is the oxygen diffusion coefficient, and C_{O_2} is the oxygen concentration. Substituting Eq. 2.20 into Eq. 2.21 and rearranging, the following expression is obtained:

$$\frac{dC_{O_2}}{dx} = -\frac{P_{O_2}\left(p_W^1, \ p_W^2\right) \cdot \frac{\Delta p_{O_2}}{l}}{D_{O_2}(C_W)}. \tag{2.22}$$

Integrating the left- and right-hand sides of Eq. 2.22 over the film thickness, and assuming that the solubilization process of oxygen into the matrix can be described by means of Henry's law, one obtains

$$P_{O_2}\left(p_W^1, \ p_W^2\right) = \frac{1}{\displaystyle\int_0^l \frac{1}{S_{O_2} \cdot D_F^{O_2}(C_W)} \cdot dx}, \tag{2.23}$$

where S_{O_2} is the oxygen-solubility coefficient. To relate $D_{O_2}(C_W)$ to C_W, the simple expression illustrated earlier to relate the water diffusion coefficient to the local water concentration (Eq. 2.15) can be used:

$$D_{O_2}(C_W) = B_1^* \cdot \exp\left(\frac{-1}{B_2^* + B_3^* \cdot C_W}\right), \tag{2.24}$$

where the B_i^* are constants without any specific physical meaning and considered fitting parameters. Substituting Eq. 2.24 into Eq. 2.23 one obtains

$$P_{O_2}\left(p_W^1, \ p_W^2\right) = \frac{1}{\displaystyle\int_0^l \frac{1}{S_{O_2} \cdot B_1^* \cdot \exp\left(\frac{-1}{B_2^* + B_3^* \cdot C_W}\right)} \cdot dx}. \tag{2.25}$$

As can be inferred from Eq. 2.25, to determine the oxygen permeability coefficient, first it is necessary to evaluate the steady-state water concentration profile. Thus, an expression similar to Eq. 2.22 can be obtained for $\frac{dC_W}{dx}$ by assuming that the water diffusion process can be described by means of Fick's model with a water diffusion coefficient dependent on the local water concentration:

$$\frac{dC_W}{dx} = -\frac{P_W\left(p_W^1, \ p_W^2\right) \cdot \frac{\Delta p_W}{l}}{D_W(C_W)}, \tag{2.26}$$

where $\frac{\Delta p_W}{l}$ is the water vapor partial pressure gradient across the film. To integrate Eq. 2.26 (i.e., to calculate the steady-state water concentration profile), both the relationship relating the water permeability to the upstream and downstream water vapor partial pressure [i.e., $P_W\left(p_W^1, \ p_W^2\right)$] and the expression relating the water diffusion coefficient to the local water concentration [i.e., $D_W(C_W)$] must be determined. As discussed earlier, whenever Fick's law holds, the water permeability coefficient is related to the water vapor partial pressure at the film sides through Eq. 2.6, even though both solubilization and diffusion processes are needed. There are several ways to approach these two distinct problems. In fact, water solubilization can be described either by taking into account the adsorbed water molecules (Eq. 2.11) or by considering only the dissolved water molecules (Eq. 2.7). Regarding the dependence of water diffusion on the local water concentration, either Eq. 2.9 or Eq. 2.15 can be used. Moreover, to calculate the water diffusion model parameters, the polymer relaxation process could either be taken or not taken into account.

A possible way to model oxygen diffusion through moderately hydrophilic film was proposed by Del Nobile et al. (2003c), who focused their attention on the particular case of oxygen diffusion through nylon film. The authors used Eq. 2.11 to describe the water solubilization process, whereas Eq. 2.15 was used to relate the water diffusion coefficient to the local water concentration. Moreover, the researchers took into account polymer relaxation to calculate the water diffusion model parameters. Under the preceding hypothesis, Eq. 2.26 can be rewritten as follows:

$$
\frac{dC_W}{dx} = -\frac{\left(\dfrac{1}{p_W^1 - p_W^2} \cdot \displaystyle\int_{C_W(p_W^2)}^{C_W(p_W^1)} B_1 \cdot \exp\left(\dfrac{-1}{B_2 + B_3 \cdot C_W}\right) \cdot dC_W \right) \cdot \dfrac{\Delta p_W}{1}}{B_1 \cdot \exp\left(\dfrac{-1}{B_2 + B_3 \cdot C_W}\right)}. \quad (2.27)
$$

The set of equations consisting of Eqs. 2.25 and 2.27 is the mathematical approach used by Del Nobile et al. (2003c) to predict the oxygen barrier properties of a nylon film. To validate the model, the authors used known model parameters.

When the water vapor partial pressure at the upstream and downstream sides of the film is equal, Eq. 2.25 becomes

$$
P_{O_2}\left(p_W^1\right) = S_{O_2} \cdot B_1 \cdot \exp\left(\frac{-1}{B_2 + B_3 \cdot C_W^*}\right), \quad (2.28)
$$

where C_W^* is the water concentration in the film.

The values obtained for the parameters appearing in Eq. 2.28 are as follows:

$$S_{O_2} \cdot B_1 = 4.39 \cdot 10^{59} \left[\frac{cm^3(STP) \cdot cm}{cm^2 \cdot s \cdot atm}\right] ; \quad B_2 = 6.28 \cdot 10^{-3} ; \quad B_3 = 1.05 \cdot 10^{-3}$$

$\left[\dfrac{g\ dry\ polymer}{g\ water}\right]$; whereas the value of $\bar{E}\%$ calculated by Del Nobile et al. (2003c) is 7.04. It is worth noting that the value of C_W^* was evaluated by means of Eq. 2.11 using data previously reported in the literature (Del Nobile et al. 2003b). As can be inferred from the $\bar{E}\%$, the ability of the proposed model to interpolate the experimental data is quite satisfactory, and it corroborates the assumptions used by the authors to derive it.

To further confirm the validity of this approach, the model was used by the same authors to also predict the oxygen permeability of the investigated film when the water activity at the upstream and downstream sides of the film is different. As reported earlier, to determine the oxygen permeability coefficient, the steady-state water concentration profile must first be evaluated. In the study of Del Nobile et al. (2003c), the steady-state water concentration profile inside the nylon film in the case of a permeation test is calculated using Eq. 2.27 with data reported in the literature (Del Nobile et al. 2003b). The authors conducted permeation tests by setting the water activity at the downstream side of the film equal to 0.35 (the water activity at the upstream side of the film ranged from 0.3 to 0.9) and predicted the oxygen permeability coefficient using Eq. 2.25. Also in this case the goodness of the model was found to be satisfactorily adequate ($\bar{E}\%$ equal to 8.45).

2.3 Multilayer Structures

In many cases flexible films intended for food-packaging applications are made by combining, laminating, or coextruding several polymeric layers. These types of film are usually referred to as multilayer systems. The prediction of the barrier properties of a multilayer film has an additional difficulty besides that encountered in predicting the mass transport properties of a single-layer structure. In fact, one must generally know the permeability and the thickness of each layer. Two different cases can be encountered: (1) where the permeability coefficient of each layer does not depend on the water vapor partial pressure at the layer boundaries and (2) where the multilayer has at least one moderately hydrophilic layer whose permeability coefficient depends on the relative humidity at its boundaries. In what follows, these two cases will be illustrated separately.

2.3.1 Constant Permeability Coefficient

This is the case where a multilayer structure is made up of polyolefin layers such as polyethylene (PE) and polypropylene (PP). The hydrophobic nature of polyolefin makes these layers insensitive to water vapor. Therefore, their permeability coefficient does not depend on the water activity at layer boundaries; consequently, the permeability coefficient of the corresponding multilayer structure does not depend on the relative humidity at the film sides.

Assuming that the barrier properties of adhesive between layers are negligible, it can be easily demonstrated that the permeability coefficient of a multilayer film is related to the permeability coefficient of its constituent layers and their thicknesses through the following equation:

$$P^{Tot} = \frac{1}{\sum\limits_{1=1}^{n} \frac{l_i}{l_{Tot}} \cdot \frac{1}{P^i}}, \qquad (2.29)$$

Table 2.2 Values of oxygen permeability of monolayer OPP, PPcast, and PPcoex and multilayer films. Tests were conducted at 23 °C and 0 % relative humidity

Film	O_2 Permeability [mol•cm/(cm^2•h•atm)]
OPP (40 µm)	$8.712 \cdot 10^{-10} \pm 5.99 \cdot 10^{-12}$
OPP (40 µm)	$8.776 \cdot 10^{-10} \pm 1.10 \cdot 10^{-12}$
OPP(40 µm)-OPP(40 µm)	$1.043 \cdot 10^{-09} \pm 4.57 \cdot 10^{-11}$
PPcoex (30 µm)	$9.870 \cdot 10^{-10} \pm 4.10 \cdot 10^{-11}$
PPcast (30 µm)	$1.394 \cdot 10^{-09} \pm 2.49 \cdot 10^{-11}$
PPcoex(30 µm)-PPcast(30 µm)	$1.369 \cdot 10^{-09} \pm 6.66 \cdot 10^{-11}$

where P^{Tot} is the multilayer film oxygen permeability, λ_{Tot} is the multilayer thickness, P^i is the ith-layer oxygen permeability, and l_i is the ith-layer thickness.

Mastromatteo and Del Nobile (2011) tested the reliability of Eq. 2.29 in the case of two different laminated multilayer structures: the first was made up of two oriented PP layers [OPP(40 µm)–OPP(40 µm)], the second one was made up of a layer of coextruded PE and PP and a layer of PP cast (PPcoex30/PPcast30). In fact, the two authors first measured the oxygen permeability of each layer and subsequently the oxygen permeability of the entire multilayer structures. The results obtained in the work are summarized in Table 2.2. The values predicted by Eq. 2.29 for the two multilayer films are 8.74×10^{-10} [mol $\cdot$ cm $\cdot$ cm$^{-2} \cdot$ h$^{-1} \cdot$ atm^{-1}] for OPP40/OPP40 and 1.194×10^{-9} [mol $\cdot$ cm $\cdot$ cm$^{-2} \cdot$ h$^{-1} \cdot$ atm^{-1}] for PPcoex30/ PPcast30. In particular, the authors used the permeability data listed in Table 2.2 for the constituent layers to predict the oxygen permeability of the corresponding hydrophobic multilayer films. $\bar{E}\%$ values calculated to quantitatively determine the goodness of prediction were 16.17 and 12.76 for OPP(40 µm)–OPP(40 µm) and PPcoex(30 µm)–PPcast(30 µm), respectively. As can be inferred from Table 2.2, the values predicted are close to the experimental ones. In fact, $\bar{E}\%$ values were both lower than 20 %. Differences between predicted and experimental values might be attributed to both material variability and the presence of adhesive between layers. To sum up, the results obtained in the work of Mastromatteo and Del Nobile (2011) suggest that Eq. 2.29 is a reliable simple tool for predicting the mass transport properties of multilayer films made up of hydrophobic polymers.

2.3.2 Relative-Humidity-Dependent Permeability Coefficient

The conventional approach to producing barrier films for food-packaging applications is to combine different material in a multilayer structure. In fact, the strategy of combining different polymers with various gas or water vapor barrier properties is also a useful means of reducing the sensitivity of layers to low molecular weight compounds and to improve multilayer performance (Buonocore et al. 2005; Chumillas et al. 2007). Generally, commercially available multilayer films have one moderately hydrophilic film, such as PET, ethylene-vinyl alcohol

(EVOH) or polyamide (PA), which serves as a barrier layer. It is usually placed between two hydrophobic films, such as polyolefin. In these cases Eq. 2.29 must take into account the dependence of film mass transport properties on upstream and downstream water vapor partial pressure. Several researchers have investigated these aspects; in what follows, the mathematical approach proposed by Del Nobile et al. (2004) to predict the water barrier properties of a multilayer structure will be illustrated. In particular, a six-layer film composed of five polyolefin layers and one layer of EVOH was used to validate the model.

The water permeability coefficient of a multilayer structure is related to the permeability and the thickness of each layer through the following expression:

$$P_W^M \left(p_W^1, \ p_W^3 \right) = \cfrac{1}{\cfrac{l_1}{l} \cdot \cfrac{1}{P_W^1} + \cfrac{l_2}{l} \cdot \cfrac{1}{P_W^2 \left(p_W^{1-2}, p_W^{2-3} \right)} + \cfrac{l_3}{l} \cdot \cfrac{1}{P_W^3}}, \qquad (2.30)$$

where $P_W^M \left(p_W^1, \ p_W^3 \right)$ is the multilayer film water permeability, p_W^1 and p_W^3 are the water vapor partial pressure at the upstream and downstream sides of the multilayer film, respectively, l_1 is the total thickness of the polyolefin layers at the upstream side of the water-sensitive layer, l_2 is the thickness of the water-sensitive layer, l_3 is the thickness of the polyolefin layers at the downstream side of the water-sensitive layer, l is the total thickness of the multilayer film, P_W^1 is the total water permeability of the polyolefin layers at the upstream side of the water-sensitive layer, $P_W^2 \times \left(p_W^{1-2}, p_W^{2-3} \right)$ is the water permeability of the water-sensitive layer, P_W^3 is total water permeability of the polyolefin layers at the downstream side of the water-sensitive layer, and p_W^{1-2} and p_W^{2-3} are the water vapor partial pressure at the upstream and downstream sides of the water-sensitive layer, respectively. As can be seen, in Eq. 2.30 there are four unknown physical quantities, P_W^1, P_W^3, p_W^{1-2}, p_W^{2-3}, and one unknown function, $P_W^2 \left(p_W^{1-2}, p_W^{2-3} \right)$. To make this expression operative, both physical quantities and function must be determined.

An approach similar to that illustrated earlier was used by Del Nobile et al. (2004) to find the $P_W^2 \left(p_W^{1-2}, p_W^{2-3} \right)$ function. In particular, the authors modified the equation proposed by Flory (1953) to describe the mixing process of a linear polymer with a low molecular weight compound to relate C_W to the water activity:

$$\ln(a_W) = \ln \left(\frac{C_W}{\rho_W + C_W} \right) + \left(1 - \frac{C_W}{\rho_W + C_W} \right)$$
$$+ \left(\chi_0 + \chi_1 \cdot a_W \right) \cdot \left(1 - \frac{C_W}{\rho_W + C_W} \right)^2. \qquad (2.31)$$

It is worth noting that Eq. 2.31 was derived by Del Nobile et al. (2004) from Flory's equation assuming that the Flory interaction parameter was a linear function of the water activity. This hypothesis is reasonable since the bound water could change the affinity between the free water and the polymeric matrix. Del Nobile et al. (2004) measured the water sorption isotherm at 25°C for the EVOH film.

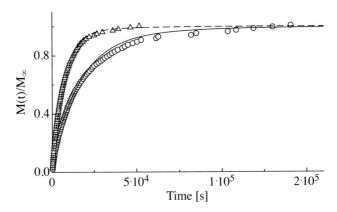

Fig. 2.7 Water sorption kinetics of EVOH film at 25 °C. (○) water activity = 0.6; (△) water activity = 0.8. The *curves* are the best fit of the model to the experimental data

By fitting Eq. 2.31 to the experimental data the following fitting parameters were obtained: $\chi_0 = 2.34$ and $\chi_1 = -0.293$. The same authors calculated the $\bar{E}\%$ value to determine the goodness of the model. A value of $\bar{E}\%$ equal to 16.6 was obtained, which is not as good as that obtained by Del Nobile et al. (2003b), even if it can be also considered quite acceptable. The authors concluded that the equation proposed by Flory (1953) with the interaction parameter dependent on water activity could be advantageously used to describe the sorption process of water into moderately hydrophilic polymers such as EVOH.

In the same work (Del Nobile et al. 2004) the water diffusion related to Brownian motion was described by means of Fick's model, with a diffusion coefficient depending on the local water concentration (Eq. 2.8). As reported in the literature (Schwartzberg 1986), for many polymer/penetrant systems the relationship between the water diffusion coefficient and the local water concentration can be expressed through an exponential type equation:

$$D_W(C_W) = \phi_1 \cdot \exp\left[\phi_2 \cdot (C_W)^4\right], \tag{2.32}$$

where ϕ_1 and ϕ_2 are constants to be regarded as fitting parameters.

To account for the water-induced polymer relaxation, the authors followed the same approach illustrated earlier. In particular, Eq. 2.16 was used to describe the elastic response of the system to the external perturbation, that is, to related K, the normalized initial water concentration at the boundaries of the film, to $C_W(0)$ and $[C_W(\infty) - C_W(0)]$. In contrast, the boundary condition relaxation rate, $\frac{d\alpha(t)}{dt}$, was expressed through Eq. 2.17. As an example, the results obtained in the work of Del Nobile et al. (2004) are shown in Fig. 2.7, where the water uptake kinetics of the EVOH film at different water activities are reported. In the same figure the curves representing the best fit of the model are also shown. Also in this case the authors simultaneously fitted the model to all the measured sorption kinetics. The following model parameters were obtained: $K_1 = 6.01 \cdot 10^{-2}$; $K_2 \left[\frac{\text{g dry polymer}}{\text{g water}}\right] = 37.3$; $\alpha_1 \left[\frac{1}{s}\right] = 2.88 \cdot 10^{-5}$;

$\alpha_2 \left[\frac{\text{g dry polymer}}{\text{g water}}\right] = 58.5$; $\phi_1 \left[\frac{\text{cm}^2}{s}\right] = 1.64 \cdot 10^{-10}$; $\phi_2 \left[\frac{\text{g dry polymer}}{\text{g water}}\right]^4 = 2.52 \cdot 10^5$; all

the values of $\bar{E}\%$ calculated in the study for each sorption kinetic curve were lower than 10. The authors were able to quantitatively express the relationship between the water diffusion coefficient and local water concentration (ϕ_1 and ϕ_2).

Finally, the EVOH water permeability was related to the water vapor pressure at the film sides through Eq. 2.6, which can be rearranged as follows using Eq. 2.32 to describe the dependence of the water diffusion coefficient on the local water concentration:

$$P_W\left(p_W^1,\ p_W^2\right) = \frac{1}{p_W^1 - p_W^2} \cdot \int_{C_W\left(p_W^2\right)}^{C_W\left(p_W^1\right)} \left\{\phi_1 \cdot \exp\left(\phi_2 \cdot C_{(W)}^4\right)\right\} \cdot dC_W. \qquad (2.33)$$

A good agreement between the measured and predicted water permeability of EVOH film by Eq. 2.33 ($\bar{E}\%$ equal to 5.74) was recorded.

The water permeability coefficient of the polyolefin layers does not depend on the external humidity. Therefore, in this case the water barrier properties of the film can be determined by a single water permeation test. The average water permeability coefficients of each polyolefin layer used by Del Nobile et al. (2004) were calculated by averaging the values measured at different water activity levels. The following values of p_W^{av} $\left[\frac{\text{g}\cdot\text{cm}}{\text{cm}^2\cdot\text{s}\cdot\text{atm}}\right]$ were obtained: $5.07 \cdot 10^{-10} \pm 7.13 \cdot 10^{-12}$, $5.53 \cdot 10^{-9} \pm 1.29 \cdot 10^{-10}$, and $1.14 \cdot 10^{-8} \pm 3.35 \cdot 10^{-10}$ for the first three polyolefin layers closest to the packed food and $1.73 \cdot 10^{-9} \pm 7.44 \cdot 10^{-11}$ for the sixth polyolefin layer that was the outer film of the multilayer system. The values of P_W^1 and P_W^3 were calculated by Del Nobile et al. (2004) by applying Eq. 2.29 to the upstream and downstream polyolefin layers, respectively. Finally, the water vapor partial pressure at the upstream and downstream sides of the water-sensitive layer (i.e., p_W^{1-2} and p_W^{2-3}, respectively) were calculated by Del Nobile et al. (2004) by numerically solving the following set of nonlinear equations:

$$\begin{cases} J = P_W^1 \cdot \dfrac{p_W^1 - p_W^{1-2}}{l_1}, \\[2mm] J = P_W^2\left(a_W^{1-2}, a_W^{2-3}\right) \cdot \dfrac{p_W^{1-2} - p_W^{2-3}}{l_2}, \\[2mm] J = P_W^3 \cdot \dfrac{p_W^{2-3} - p_W^3}{l_3}. \end{cases} \qquad (2.34)$$

The value of $\bar{E}\%$ relative to the prediction of the water permeability coefficient of the multilayer film measured by Del Nobile et al. (2004) as a function of the water activity at the upstream side of the film (the water activity at the downstream side of the film was set to zero) is 7.05 %, suggesting that the approach proposed to predict the water permeability coefficient of a multilayer film from a knowledge of the water transport properties of each constituent layer can be successfully utilized.

A much simpler mathematical model was proposed by Matromatteo and Del Nobile (2011) to predict the oxygen permeability coefficient of multilayer films with one water-sensitive layer. It is based on the observation that generally, water-sensitive multilayer films are made of hydrophobic polymers, such as polyolefin, and moderately hydrophilic polymers, such as PA and EVOH; this feature makes the water concentration gradient across moderately hydrophilic polymers very small, making the assumption of constant water concentration across this type of layers reasonable. According to this assumption, Eq. 2.30 becomes

$$P_{O_2}^M \left(p_W^1, \ p_W^3 \right) = \cfrac{1}{\cfrac{l_1}{1} \cdot \cfrac{1}{P_{O_2}^1} + \cfrac{l_2}{1} \cdot \cfrac{1}{P_{O_2}^2 \left(\overline{p_W^2} \right)} + \cfrac{l_3}{1} \cdot \cfrac{1}{P_{O_2}^3}}, \qquad (2.35)$$

where $P_{O_2}^M \left(p_W^1, \ p_W^3 \right)$ is the multilayer film oxygen permeability, $P_{O_2}^1$ is the total oxygen permeability of the polyolefin layers at the upstream side of the water-sensitive layer, $P_{O_2}^2 \left(\overline{p_W^2} \right)$ is the oxygen permeability of the water-sensitive layer, $P_{O_2}^3$ is the total oxygen permeability of the polyolefin layers at the downstream side of the water-sensitive layer, and $\overline{p_W^2}$ is the average water vapor partial pressure across the water-sensitive layer.

Mastromatteo and Del Nobile (2011) tested the model with two water-sensitive multilayer structures: PET/PPcast and OPA/OPA/PE. Moreover, the authors ran two different permeation tests: the same water vapor partial pressure at the film sides and different water vapor partial pressure at the film sides. Regarding the dependence of oxygen permeability of the single layer on the average water vapor partial pressure, the authors used Eqs. 2.1 and 2.2 for PET and OPA, respectively (Fig. 2.1). The oxygen permeability of PPcast50 and PE110 films, measured at 23 °C and 0 %RH, were found to be equal to $2.09 \cdot 10^{-09}$ and $3.97 \cdot 10^{-09} \frac{mol \cdot cm}{cm^2 \cdot h \cdot atm}$, respectively. Figure 2.8 shows the oxygen permeability coefficient, as measured by Mastromatteo and Del Nobile (2011), plotted as a function of water vapor partial pressure for PET12/PPcast50 and OPA15/OPA15/PE110 multilayer films, respectively. In this case the permeability tests were run setting the same water vapor partial pressure at the film sides. The curves shown in each graph were predicted by means of Eq. 2.35. It is worth noting that both of the multilayers used by the authors are bilayer structures. Therefore, Eq. 2.35 can be simplified by taking into account only one polyolefin layer. The authors calculated the $\bar{E}\%$ for both tested multilayer structures and obtained 14 and 20 for PET12/PPcast50 and OPA15/OPA15/PE110, respectively. The $\bar{E}\%$ values calculated in the work were not as low as that found by Del Nobile et al. (2004). However, considering the simplicity of the model proposed, the $\bar{E}\%$ found still seems to be quite acceptable.

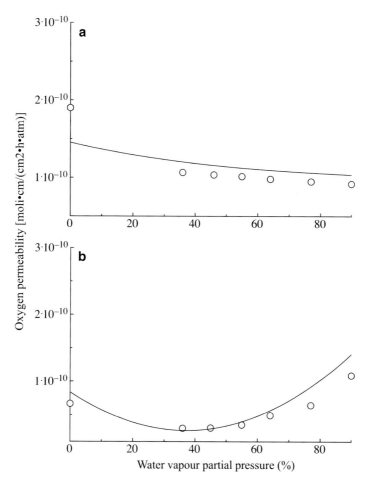

Fig. 2.8 Oxygen permeability coefficient plotted as function of water vapor partial pressure for PET12/PPcast50 (**a**) and OPA15/OPA15/PE110 multilayer film (**b**). The *curves* shown in each graph were predicted using Eq. 2.35

Equation 2.35 was further tested by the same authors by running permeation tests simulating the real working conditions of a flexible packaging film. This second step was run maintaining a constant relative humidity gradient across the multilayer for the entire period of the permeation test. In other words, different values of water vapor partial pressure were set at the film sides. Figure 2.9a shows the PET12/PPcast50 multilayer oxygen permeability plotted as a function of water vapor partial pressure at the PPcast side. As one would expect, when the relative humidity is changed at the PPcast side, the average relative humidity across PET12 film is practically constant. Therefore, the PET12/PPcast50 oxygen permeability does not change much. The curve shown in Fig. 2.9a was predicted by means of Eq. 2.35; the average water vapor partial pressure across the PET12 layer was set

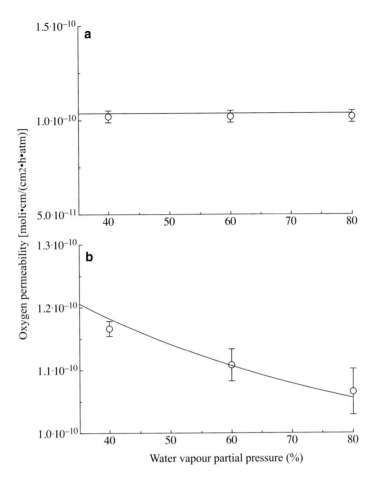

Fig. 2.9 Oxygen permeability coefficient plotted as function of water vapor partial pressure at (**a**) PPcast side and (**b**) PET side for PET12/PPcast50 multilayer film. The *curves* shown in the figure were predicted using Eq. 2.35.

equal to the water vapor partial pressure at the PET12 side. The value of $\bar{E}\%$ found by the authors was 1.7, which is an adequate value. Figure 2.9b shows the oxygen permeability of PET12/PPcast50 multilayer film plotted as a function of the water vapor partial pressure at the PET12 side. In this case a decrease in the oxygen permeability with an increase in the relative humidity was detected. This is because the average water vapor partial pressure across the PET12 layer is practically equal to that at the PET12 film side; consequently, a decrease in the latter coincides with a decrease in the former. According to the data shown in Fig. 2.1, a decrease in the water vapor partial pressure across PET12 causes a decrease in its oxygen permeability and, consequently, a decrease in the oxygen permeability of the PET12/PPcast50 multilayer film. Also in this case the prediction of Eq. 2.35 is shown in the figure.

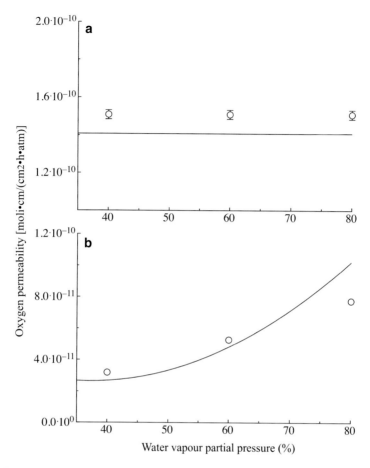

Fig. 2.10 Oxygen permeability coefficient plotted as function of water vapor partial pressure at (**a**) PE side and (**b**) OPA side for the OPA15/OPA15/PE110 multilayer film. The *curves* shown in the figure were predicted by using Eq. 2.35

The calculated $\bar{E}\%$ value was 0.79, which seems to be excellent also considering the simplicity of the model used to predict the gas permeability.

Figure 2.10 shows the OPA15/OPA15/PE110 multilayer film oxygen permeability plotted as a function of relative humidity at the PE110 and OPA sides, respectively (Mastromatteo and Del Nobile 2011). When the water vapor partial pressure at the water-sensitive layer side is kept constant, the multilayer oxygen permeability does not change much, whereas if the relative humidity at the water-sensitive layer side is changed, then the multilayer oxygen permeability also changes. The curves shown in both figures were predicted using Eq. 2.35. The $\bar{E}\%$ values were 6.7 and 18.9 for the data shown in Fig. 2.10. Both $\bar{E}\%$ values suggest that the model proposed by Mastromatteo and Del Nobile (2011) is a good compromise between accuracy of prediction and model simplicity.

2.4   Food Deterioration Modeling

The deterioration of packaged food during storage is the other element of a mechanistic SL model. In fact, several equations have been reported in the literature to describe the detrimental phenomena responsible for food quality loss during storage. For instance, Eq. 1.5 was proposed by Labuza (1971) to relate the extent of oxidation of a dry food to the extent of oxidation reaction, the water vapor partial pressure, and the oxygen partial pressure in the package headspace. Other examples are water sorption isotherms, useful for indicating the loss of crispiness in cereal-based dry products during storage. In this section, modeling the respiration rate of minimally processed produce will be discussed since it is related to fresh produce deterioration.

Minimal processing of fresh produce generally consists in washing, cutting, treating with sanitizing agents, packaging, and storing under refrigerated conditions (McKellar et al. 2004). Peeling and cutting in particular can cause damage to vegetable tissues, such as breakage of the cell walls and loss of intracellular substances and enzymes. Minimal processing gives additional value to fruits and vegetables in terms of convenience and time saving; however, it causes rapid deterioration by increasing the respiration rate, the transpiration, the enzymatic activity, and microbial proliferation (Nguyen-the and Carlin 1994), thereby reducing the SL of the packaged product to a few days. The respiration rate of minimally processed crops, which is an index of their quality decay, increases 1.2- to 7.0-fold depending on the produce, cutting grade, and temperature (Ahrenainen 1996).

Fresh-cut produce is usually stored under MAP. Low O_2 and elevated CO_2 atmospheres, together with low storage temperature, reduce the product respiration rate (Watada et al. 1996), thereby limiting losses in fresh weight and in dry matter (Böttcher et al. 2003). Such modified atmospheres can be achieved either actively, by filling the packages with specific gas mixtures, or passively, as a consequence of the respiratory activity of the packaged produce, O_2 consumption and CO_2 production, and gas transfer through the packaged film (Jacxsens et al. 1999). A proper combination of product characteristics and film permeability results in the evolution of an appropriate atmosphere within packages (Smith et al. 1987). In addition to the permeability, the gas selectivity of a film must also be considered. The equilibrium gas concentrations of packaged produce also depend on packaged product weight, respiring surface area, and storage temperature.

As the quality loss of minimally processed fruit is mainly determined by physiological events (i.e., increased respiration), it is of paramount importance to create conditions in the packages that could slow down the respiration process. Several models are available in the literature to describe the produce respiration rate, and most of them are based on the Michaelis–Menten equation. Chevillotte (1973) introduced Michaelis–Menten kinetics. Lee et al. (1991) included uncompetitive inhibition by CO_2 and tested the model on cut broccoli. Peppelenbos and Van't Leven (1996) proposed four types of inhibition to model the influence of CO_2 levels on O_2 consumption of fruits and vegetables, as compared to cases where there is no influence of CO_2 on the respiration rate (i.e., no inhibition). They introduced an equation to describe the O_2 consumption rate as inhibited by CO_2 both in a competitive and in an uncompetitive way. Hertog et al. (1998) described and discussed multiple facets of the formulation for the combined types of inhibition of the O_2 consumption rate by CO_2 concentration. The related mathematical equations based on the Michaelis–Menten kinetic for no-inhibition, competitive inhibition, and uncompetitive inhibition are presented in the following expressions:

$$r_{O_2} = \frac{Vm_{O_2} \cdot [O_2]}{Km_{O_2} + [O_2]}, \tag{2.36}$$

$$r_{O_2} = \frac{Vm_{O_2} \cdot [O_2]}{Km_{O_2} \cdot \left\{1 + \frac{[CO_2]}{Kmc_{CO_2}}\right\} + [O_2]}, \tag{2.37}$$

$$r_{O_2} = \frac{Vm_{O_2} \cdot [O_2]}{Km_{O_2} + [O_2] \cdot \left\{1 + \frac{[CO_2]}{Kmu_{CO_2}}\right\}}, \tag{2.38}$$

where r_{O_2} is the produce oxygen consumption rate, $[O_2]$ and $[CO_2]$ are the oxygen and carbon dioxide package headspace concentrations, respectively, Vm_{O_2} is the maximum O_2 consumption rate, Km_{O_2} is the Michaelis constant for oxygen consumption, Kmc_{CO_2} is the Michaelis constant for the competitive inhibition of the oxygen consumption rate by carbon dioxide, and Kmu_{CO_2} is the Michaelis constant for the uncompetitive inhibition of the oxygen consumption rate by carbon dioxide.

As an alternative to the models described above, an empirical approach was proposed by Del Nobile et al. (2006), where it is assumed that the oxygen consumption rate depends linearly on the headspace oxygen concentration:

$$r_{O_2} = \Phi \cdot [O_2], \tag{2.39}$$

where Φ is the kinetic constant of the entire respiration process. The parameter Φ depends on temperature and, in some cases, on the carbon dioxide concentration. In the work of Del Nobile et al. (2006) it was assumed that Φ depends on the carbon dioxide concentration through an exponential-type expression:

$$\Phi = \Phi_1 \cdot \exp\{-\Phi_2 \cdot [CO_2]\}, \tag{2.40}$$

where Φ_1 is the preexponential term and is the maximum oxygen consumption rate, Φ_2 is the exponential factor and accounts for carbon-dioxide-induced respiration inhibition. Substituting Eq. 2.40 into Eq. 2.39 one obtains

$$r_{O_2} = \Phi_1 \cdot \exp\{-\Phi_2 \cdot [CO_2]\} \cdot [O_2]. \tag{2.41}$$

It is generally assumed that moles of carbon dioxide produced are equal to those of oxygen consumed (Lee et al. 1991; Makino et al. 1997). To describe the carbon dioxide consumption rate, Del Nobile et al. (2006) assumed that the ratio between carbon dioxide produced and oxygen consumed that corresponds to the respiratory quotient (RQ) is constant, but not necessarily equal to 1 (1 is the theoretical value); a change in the RQ can indicate the nature of the respiratory substrate and whether anaerobic respiration is occurring (Joles et al. 1994):

$$r_{CO_2} = \Psi_1 \cdot \{\Phi_1 \cdot \exp\{-\Phi_2 \cdot [CO_2]\} \cdot [O_2]\}, \tag{2.42}$$

where r_{CO_2} is the carbon dioxide production rate, and Ψ_1 is the ratio between the moles of carbon dioxide produced and the moles of oxygen consumed.

To describe the time course during storage of oxygen and carbon dioxide concentration in a package, the mass balance on these two substances in the package headspace is written as

$$\frac{d(n_{O_2}(t))}{dt} = S \cdot P_{O_2} \cdot \frac{\left[p_{O_2}^{est} - \frac{n_{O_2}(t) \cdot R \cdot T}{V_{st}}\right]}{1} +$$
$$- mp \cdot 4.615 \cdot 10^{-6} \cdot \{\Phi_1 \cdot [O_2] \cdot \exp\{-\Phi_2 \cdot [CO_2]\}\}, \tag{2.43}$$

$$\frac{d(n_{CO_2}(t))}{dt} = S \cdot P_{O_2} \cdot \frac{\left[p_{CO_2}^{est} - \frac{n_{CO_2}(t) \cdot R \cdot T}{V_{st}}\right]}{1}$$
$$+ mp \cdot 4.615 \cdot 10^{-6} \cdot \{\Psi_1 \cdot \{\Phi_1 \cdot [O_2] \cdot \exp\{-\Phi_2 \cdot [CO_2]\}\}\}, \tag{2.44}$$

where $n_{O_2}(t)$ is the mole of oxygen in the package headspace at time t, S is the package surface area, P_{O_2} is the package oxygen permeability, $p_{O_2}^{est}$ is the external oxygen partial pressure, mp is the mass of the packaged produce, V_{st} is the volume of the package headspace, T is the temperature, R is the universal gas constant, $n_{CO_2}(t)$ is the mole of carbon dioxide in the package headspace at time t, P_{CO_2} is the package carbon dioxide permeability, and $p_{CO_2}^{est}$ is the external carbon dioxide partial pressure.

Equations 2.43 and 2.44 are a set of two ordinary differential equations, which can be numerically solved with the appropriate initial conditions providing the evolution during storage of the headspace oxygen and carbon dioxide concentration.

Del Nobile et al. (2007) tested both the fitting and predictive ability of the model illustrated earlier with three different fresh processed fruits: prickly pear, banana, and kiwifruit. The authors used two different films to validate the model: (1) a laminated high barrier film made up of a PE layer, aluminum foil, and a PET layer, henceforth referred to as laminated film; (2) a coextruded high-gas-permeability polyolefin film, henceforth referred to as coextruded film. As an example, Fig. 2.11a shows the variation in headspace gas composition during storage as measured by Del Nobile et al. (2007) for kiwifruit. The tests were run using the laminated film,

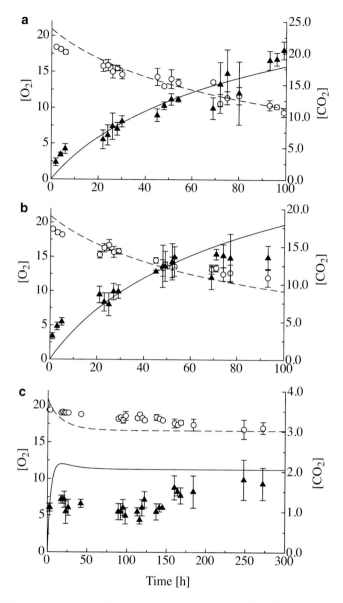

Fig. 2.11 Headspace gas composition during storage in packages of kiwifruit. Data were obtained in a laminated high barrier film made up of PE, aluminum, and PET. The *curves* were obtained by fitting the set of ordinary differential Eqs. 2.43 and 2.44 to the experimental data relative to different batches of fruit [(**a**) first batch (**b**) second batch, and (**c**) third batch]

which was considered by the authors to be impermeable to low molecular weight compounds. In fact, the carbon dioxide permeation test run on a package made up of laminated film revealed no passage of CO_2 through the package during 28 days of monitoring. As expected, the graph shows a decrease in the headspace oxygen concentration along with an increase in the headspace carbon dioxide concentration. The curves shown in the figure were obtained by Del Nobile et al. (2007) by fitting the set of ordinary differential Eqs. 2.43 and 2.44 to the experimental data. The following fitting parameter values were obtained for kiwi: 0.44 for Φ_1 (mL/Kg*h), 4.21×10^{-2} for Φ_2, and 1.97 for Ψ. It is worth noting that Del Nobile et al. (2007) simultaneously fitted the set of ordinary differential Eqs. 2.43 and 2.44 to both the oxygen and carbon dioxide concentration data. The ability of Eqs. 2.43 and 2.44 to fit the experimental data was evaluated by the $\bar{E}\%$. In the specific case of kiwi data a value of $\bar{E}\%$ equal to 11.1 was obtained, thus demonstrating that despite the simplicity of the model proposed by Del Nobile et al. (2007), its ability to interpolate the experimental data is satisfactory.

To test the predictive ability of Eqs. 2.43 and 2.44, a second set of tests was run. In particular, Del Nobile et al. (2007) used a different batch of fruits and the same laminated film used to run the first set of tests. Figure 2.11b shows the evolution of the headspace gas composition during storage for this second set of tests. The curves shown in the preceding figure were predicted by means of Eqs. 2.43 and 2.44, using the calculated values of the fitting parameters. The effectiveness of prediction was slightly lower compared to the first test ($\bar{E}\% = 13.4$). However, considering the variability associated with the packaged produce, the results obtained can be considered fairly acceptable.

To make the validation test even more rigorous, a third set of tests was run by Del Nobile et al. (2007). In particular, another batch of fruit was packaged using the coextruded film, which is characterized by a high gas permeability compared to the laminated film. Figure 2.11c shows the evolution during storage of the headspace gas composition during storage for the third set of tests. Also in this case, the curves shown in the figure were predicted by means of Eqs. 2.43 and 2.44, using the fitting parameters reported earlier. Despite the deviation of the experimental data from the model prediction ($\bar{E}\% = 39.7$), the model proposed was considered reasonable by Del Nobile et al. (2007) in light of the changeability associated with fresh-cut produce and the error associated with the measurement of the package gas permeability.

References

Ahrenainen R (1996) New approaches in improving the shelf life of minimally processed fruit and vegetables. Trends Food Sci Technol 7:179–187

Boquet R, Chirifie J, Iglesias HA (1978) Equations for fitting water sorption isotherms of foods. II. Evaluation of various two-parameter models. J Food Technol 13:319–327

Böttcher H, Günther I, Kabelitz L (2003) Physiological postharvest responses of Common Saint-John's wort herbs (*Hypericum perforatum* L.). Postharvest Biol Technol 29:342–350

Buonocore GG, Conte A, Corbo MR, Sinigaglia M, Del Nobile MA (2005) Mono and multilayer active films containing lysozyme as antimicrobial agent. Innov Food Sci Emerg Technol 6:459–464

Chevillotte P (1973) Relation between the reaction cytochrome oxidaseoxygen and oxygen uptake in cells in vivo, the role of diffusion. J Theor Biol 39:277–295

Chumillas MR, Belisario Y, Iguaz A, Lopez A (2007) Quality and shelf life of orange juice aseptically packaged in PET bottles. J Food Eng 79:234–242

Del Nobile MA, Mensitieri G, Netti PA, Nicolais L (1994) Anomalous diffusion in poly-ether-ether-ketone. Chem Eng 49:633–644

Del Nobile MA, Mensitieri G, Manfredi C, Arpaia A, Nicolais L (1996a) Low molecular weight molecules diffusion in advanced polymers for food packaging applications. Polym Adv Technol 7:409–417

Del Nobile MA, Mensitieri G, Nicolais L (1996b) Effect of chemical composition on gas transport properties of ethylene based ionomers. Polym Int 41:73–78

Del Nobile MA, Mensitieri G, Nicolais L, Masi P (1997a) The influence of the thermal history on the shelf life of carbonated beverages bottled in plastic containers. J Food Eng 34:1–13

Del Nobile MA, Mensitieri G, Ho LH, Huang SJ, Nicolais L (1997b) Moisture transport properties of a degradable nylon for food packaging. Packag Technol Sci 10:311–330

Del Nobile MA, Fava P, Piergiovanni L (2002) Water transport properties of cellophane flexible films intended for food packaging applications. J Food Eng 53:295–300

Del Nobile MA, Ambrosino ML, Sacchi R, Masi P (2003a) Design of plastic bottles for packaging of virgin olive oil. J Food Sci 68:170–175

Del Nobile MA, Buonocore GG, Altieri C, Battaglia G, Nicolais L (2003b) Modeling the water barrier properties of nylon film intended for food packaging applications. J Food Sci 68:1334–1340

Del Nobile MA, Buonocore GG, La Notte E, Nicolais L (2003c) Modeling the oxygen barrier properties of nylon film intended for food packaging applications. J Food Sci 68:2017–2021

Del Nobile MA, Buonocore GG, Dainelli D, Nicolais L (2004) A new approach to predict the water transport properties of multilayer films intended for food-packaging applications. J Food Sci 69:85–90

Del Nobile MA, Baiano A, Benedetto A, Massignan L (2006) Respiration rate of minimally processed lettuce as affected by packaging. J Food Eng 74:60–69

Del Nobile MA, Licciardello F, Scrocco C, Muratore G, Zappa M (2007) Design of plastic packages for minimally processed fruits. J Food Eng 79:217–224

Flory PJ (1953) Principles of polymer chemistry. Cornell University Press, Ithaca, pp 495–512

Fujita H (1961) Diffusion in polymer-diluent systems. Fortschr Hochpolym-Forsch 3:1–47

Gavara R, Hernandez RJ (1994) Effect of water on the transport of oxygen through nylon-6 films. J Polym Sci 32:2375–2382

Hertog MLATM, Peppelenbos HW, Evelo RG, Tijskens LMM (1998) A dynamic and generic model of gas exchange of respiring produce: the effects of oxygen, carbon dioxide and temperature. Postharvest Biol Technol 14:335–349

Jacxsens L, Devlieghere F, Debevere J (1999) Validation of a systematic approach to design equilibrium modified atmosphere packages for fresh-cut produce. Lebensm Wiss Technol 32:425–432

Joles DW, Cameron AC, Shirazi A, Petracek PD, Beaudry RM (1994) Modified-atmosphere packaging of "Heritage" red raspberry fruit: respiratory response to reduced oxygen, enhanced carbon dioxide and temperature. J Am Soc Horticult Sci 119:540–545

Labuza TP (1971) Kinetics of lipid oxidation in foods. CRC Crit Rev Food Technol 2:355–405

Lee DS, Haggar PE, Lee J, Yam KL (1991) Model for fresh produce respiration in modified atmospheres based on principles of enzyme kinetics. J Food Sci 56:1580–1585

Long FA, Richman D (1960) Concentration gradients for diffusion of vapours in glassy polymers and their relation to time dependent diffusion phenomena. J Am Chem Soc 82:513–519

Makino Y, Iwasaki K, Takashi H (1997) Application of transition state theory in model development for temperature dependence of respiration of fresh produce. J Agric Eng Res 67:47–59

Mastromatteo M, Del Nobile MA (2011) A simple model to predict the oxygen transport properties of multilayer films. J Food Eng 102:170–176

McKellar RC, Odumeru J, Zhou T, Harrison A, Mercer DG, Young JC et al (2004) Influence of a commercial warm chlorinated water treatment and packaging on the shelf-life of ready-to-use lettuce. Food Res Int 37:343–354

Netti PA, Del Nobile MA, Mensitieri G, Ambrosio L, Nicolais L (1996) Water transport in hyaluronic acid esters. J Bioact Compat Polym 11:312–327

Nguyen-the C, Carlin F (1994) The microbiology of minimally processed fresh fruits and vegetables. Crit Rev Food Sci Nutr 34:371–401

Peppelenbos HW, Van't Leven J (1996) Evaluation of four types of inhibition for modelling the influence of carbon dioxide on oxygen consumption of fruits and vegetables. Postharvest Biol Technol 7:27–40

Quast DG, Karel M (1972) Computer simulation of storage life of foods undergoing spoilage by two interacting mechanisms. J Food Sci 37:679–683

Schwartzberg HG (1986) Modelling of gas and vapor transport trough hydrophilic films. In: Mathlouthi M (ed) Food packaging and preservation. Elsevier, New York, pp 115–135

Smith S, Geeson J, Stow J (1987) Production of modified atmosphere in deciduous fruits by the use of films and coatings. Horticult Sci 22:772–776

Talasila PC, Cameron AC (1997) Prediction equations for gases in flexible modified-atmosphere packages of respiring produce are different than those for rigid packages. J Food Sci 62:926–930

Watada AE, Ko NP, Minott DA (1996) Factors affecting quality of fresh-cut horticultural products. Postharvest Biol Technol 9:115–125

Chapter 3
Mechanistic Models for Shelf Life Prediction

3.1 Introduction

In this chapter the models illustrated and discussed in previous chapters to describe both the package barrier properties and the food deterioration processes are combined to derive mechanistic models. There are several ways to cluster the models, and the way the mass flux is calculated is one of the possible modes of gathering them. In fact, there are two methods for calculating the mass flux of low molecular weight compounds through a package: (1) under pseudo-steady-state conditions, i.e., the mass exchanged during the transient state can be neglected; (2) under unsteady-state conditions, i.e., the mass exchanged during the transient state must be taken into account. These approaches are profoundly different, and consequently the way to derive the shelf life (SL) model is different. Therefore, they will be presented and discussed separately in what follows.

3.2 Pseudo-Steady-State Conditions

The simplest way to calculate the amount of low molecular weight compound exchanged between the inside and outside of a package assumes that the pseudo-steady-state conditions are obeyed. As reported in the literature, this is a quite reasonable assumption for flexible packages (Conte et al. 2009; Del Nobile et al. 2008a, b, 2009; Lucera et al. 2010; Rai et al. 2008). In these cases the mass flux can be calculated using Eq. 4 of part I. As was reported earlier, one of the most important approximations made using Eq. 4 of part I is that the mass exchanged during the transient state is neglected. As will be shown subsequently, this approximation is reasonable for flexible packaging but not for rigid containers such as plastic bottles.

M.A. Del Nobile and A. Conte, *Packaging for Food Preservation*,
Food Engineering Series, DOI 10.1007/978-1-4614-7684-9_3,
© Springer Science+Business Media New York 2013

Two examples of a SL model derived according to a mechanistic approach will be discussed separately in what follows. In the first example, the packaged-food quality is expressed by means of a single quality index, whereas in the second one, two different quality indices are used to describe the packaged-food quality.

3.2.1 Cereal-Based Dry Products

The quality of cereal-based dry products, such as biscuits and crackers, is strictly related to their water content. Usually, when these products are packaged, the water activity inside the package is as low as possible. Afterward, during distribution and storage, water molecules permeate through the package due to the difference between the water activity inside and that outside the package, leading to an increase in the internal water activity. This causes an increase in water content and, consequently, a decrease in product quality.

The decay kinetic of these types of food depends on several factors, of which the most important are the characteristic of the packed product (i.e., the affinity between food and water) and the water barrier properties of the package. An experimental evaluation of the optimal packaging conditions (in terms of package and food properties) is usually avoided because it is time consuming; instead, a rough estimation is generally made by means of empirical methods. As an alternative to this latter methodology, the optimal packaging conditions can be determined by means of mathematical models that are able to predict the SL of the packed product. This approach was first introduced by Heiss (1958), and later researchers proposed various mathematical models in which the moisture sensitivity of the food and the package performances were combined in different ways (Iglesias et al. 1979; Labuza and Contreras-Medellin 1981; Tubert and Iglesias 1985; Fava et al. 2000).

Del Nobile et al. (2003b) proposed a mechanistic mathematical model that was able to predict the SL of cereal-based dry products by taking into account the dependence of the water barrier properties of the packaging film on the water activity inside and outside the package. In particular, the authors first described the sole packaged-food quality index by means of the water sorption isotherm, then expressed the water permeability coefficient of the package as a function of the water vapor partial pressure at the film sides, and finally combined the preceding information using the water mass balance equation inside the package. In what follows, this mechanistic model will be illustrated and discussed in detail.

The water sorption isotherm of a given food is generally used to quantitatively determine its affinity with water. In the particular case of foods whose quality depends only on the amount of absorbed water, such as cereal-based dry products, the sorption isotherm is also used to relate the quality of the product to the water activity in the package headspace. Several models have been reported in the literature to describe the water sorption isotherm of foods (Bell and Labuza 2000). In this case the Guggenheim-Anderson-de Boer (GAB) equation,

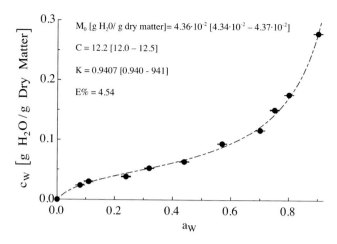

M_0 [g H_2O/ g dry matter]= $4.36 \cdot 10^{-2}$ [$4.34 \cdot 10^{-2} - 4.37 \cdot 10^{-2}$]

$C = 12.2$ [$12.0 - 12.5$]

$K = 0.9407$ [$0.940 - 941$]

$E\% = 4.54$

Fig. 3.1 Equilibrium water concentration plotted as function of water activity (a_w) for salted crackers. The curve was obtained by fitting Eq. 3.1 to the experimental data

which was recommended by the COST Project to fit moisture sorption isotherms (Wolf et al. 1985) and had already been used to successfully describe the water sorption isotherms of several foods (Saravacos 1986), has been used to fit the experimental data:

$$c_W = \frac{M_0 \cdot \Lambda \cdot \Gamma \cdot a_W}{(1 - \Gamma \cdot a_W) \cdot (1 - \Gamma \cdot a_W + \Lambda \cdot \Gamma \cdot a_W)}, \qquad (3.1)$$

where c_w is the food water concentration at equilibrium [calculated according to the following expression: $c_W = \frac{(w_f - w_i) + w_i \cdot \frac{\%H_2O}{100}}{w_i \cdot \frac{100 - \%H_2O}{100}}$, where w_f is the final product's weight, w_i the initial product's weight, $\%H_2O$ the initial moisture content (g/100 g) of the food product on a wet basis], M_O, Γ, and Λ are constants to be regarded as fitting parameters, and a_w is the water activity. Del Nobile et al. (2003b) used Eq. 3.1 to interpolate data of three different cereal-based products: unsalted and salted crackers and dry cookies. As an example, in Fig. 3.1 c_w is plotted as a function of a_w for salted crackers. The curve shown in the same figure are the best fit of Eq. 3.1 to the experimental data; results from fitting are also reported in the figure, along with $\bar{E}\%$, as calculated by Del Nobile et al. (2003b). As can be inferred, the GAB equation successfully interpolated the experimental data.

As reported earlier, multilayer structures are the most widely used flexible films for food-packaging applications. Del Nobile et al. (2003b) proposed a bilayer structure to package a cereal-based dry product made of a polyolefinic and a polyamide layer; the former layer is put in contact with the packaged food and used as a sealable layer, whereas the polyamide layer was used as a barrier film.

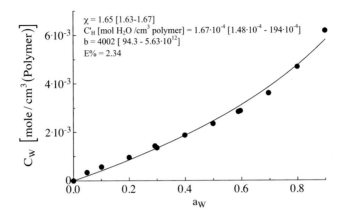

Fig. 3.2 Water sorption isotherms at 25°C of a polyamide. The curve is the best fit of Eq. 2.11 to the experimental data. The model's parameters are also reported, along with the relative E% value

Equation 2.30, used for a multilayer system, can be rearranged for a bilayer film as follows:

$$P_W^M\left(a_W^1, a_W^2\right) = \cfrac{1}{\cfrac{\ell_1}{\ell} \cdot \cfrac{1}{P_W^1\left(a_W^1, a_W^{1-2}\right)} + \cfrac{\ell_2}{\ell} \cdot \cfrac{1}{P_W^2}},$$
(3.2)

where a_W^1 and a_W^2 are the water activity at the upstream and downstream sides of the film, respectively, a_W^{1-2} is the water activity at the interface between the two layers, and $P_W^1\left(a_W^1, a_W^{1-2}\right)$ is the water permeability of the polyamide film. The water activity was calculated as the ratio between the water vapor partial pressure and the water vapor pressure at equilibrium. The authors followed a mechanistic approach similar to that reported earlier to express the polyamide water permeability as a function of water activity at the film sides. Therefore, the function $P_W^1\left(a_W^1, a_W^{1-2}\right)$ was expressed using Eq. 2.6, which can be rewritten as follows by substituting the water vapor partial pressure with the water activity:

$$P_W^1\left(a_W^1, a_W^{1-2}\right) = \frac{\int_{C_W^{1-2}\left(a_W^{1-2}\right)}^{C_W^1\left(a_W^1\right)} D_W(C_W) \cdot dC_W}{P_W^* \cdot \left(a_W^1 - a_W^{1-2}\right)}.$$
(3.3)

The polyamide water isotherm was described using Eq. 2.11. The dissolved and adsorbed water were expressed by means of Eqs. 2.12 and 2.13, respectively. For example, Fig. 3.2 shows the water sorption isotherms at 25 °C of a polyamide as measured by Del Nobile et al. (2003b). The curves shown in the figure are the best fit of Eq. 2.11 to the experimental data; the model parameters calculated are also reported in the figure, along with the relative $\bar{E}\%$ value. Also in this case there is a good agreement between the predicted and experimental data, suggesting that the sorption process of water in moderately hydrophilic polymers can indeed be described in terms of the dual sorption mechanism.

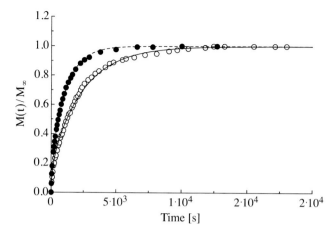

Fig. 3.3 Water sorption isotherms at 25 °C of a polyamide. ($\bigcirc$) from $a_w = 0$ to $a_w = 0.05$; ($\bullet$) from $a_w = 0.3$ to $a_w = 0.4$. The *curves* are the best fit of Eq. 2.10 to the experimental data

In the same work by Del Nobile et al. (2003b) water diffusion through moderately hydrophilic polymers was described using Fick's model with a diffusion coefficient dependent on the local water concentration (Eq. 2.8). The authors used Eq. 2.15 to express this dependence. To calculate the parameters appearing in Eq. 2.15, the fitting of Eq. 2.10 to the water sorption kinetic data was carried out. Figure 3.3 shows two examples of sorption kinetic curves of water in polyamide film, where $\frac{M(t)}{M_\infty}$ is plotted as a function of time. In the same figure the best fit of Eq. 2.10 to the water uptake kinetic data is also reported. To quantitatively determine the ability of Eq. 2.10 to interpolate the water uptake data, the authors calculated the $\bar{E}\%$ values for all the sorption tests conducted, concluding that the water diffusion coefficient could indeed be considered constant during each sorption test.

As illustrated earlier, once the water diffusion coefficient is calculated for each of the water sorption kinetics, its value is associated to the average water concentration of the corresponding water sorption kinetic. In Fig. 3.4 the water diffusion coefficient, calculated by fitting Eq. 2.10 to the water sorption kinetic data, is reported as a function of the average water concentration. In the same figure the best fit of Eq. 2.15 to the data is also reported, along with calculated model parameters and relative $\bar{E}\%$ values. Also in this case the agreement between the model predictions and the experimental data seems to be satisfactory.

Finally, the water activity at the polyamide/polyolefin interface (i.e., a_W^{1-2}) was calculated by numerically solving the following set of nonlinear equations:

$$
\begin{cases}
J = P_W^1\left(a_W^1, a_W^{1-2}\right) \cdot \dfrac{a_W^1 - a_W^{1-2}}{\ell_1}, \\[2ex]
J = P_W^2 \cdot \dfrac{a_W^{1-2} - a_W^2}{\ell_2}.
\end{cases}
\tag{3.4}
$$

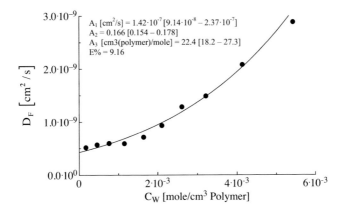

Fig. 3.4 Water diffusion coefficient reported as a function of average water concentration. The *curve* is the best fit of Eq. 2.15 to the data, along calculated model's parameters and relative E%

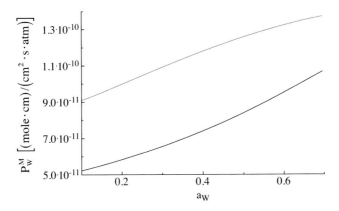

Fig. 3.5 Water permeability coefficients of two different bilayer structures plotted as function of water activity at upstream side of polyamide film. Each structure was intended as a bilayer structure made up of a layer of polyethylene (10 μm thick) and a polyamide layer

Figure 3.5 shows the calculated permeability coefficients of two different bilayer structures plotted against water activity at the upstream side of the polyamide film. Each structure was intended as a bilayer structure composed of a layer of polyethylene (10 μm thick) with a water permeability coefficient equal to $30.14 \cdot 10^{-12} \; \frac{mole \cdot cm}{cm^2 \cdot s \cdot atm}$, as reported in the literature (Myers et al. 1961), and a polyamide layer. In the work of Del Nobile et al. (2003) the permeability coefficients were calculated by considering the water activity at the downstream side of the bilayer films (i.e., the polyethylene side) equal to zero and applying Eq. 3.2 with the calculated fitting parameters. It is worth noting that the curves shown in Fig. 3.5 represent the water permeability coefficient of films as measured by a permeability test where the polyamide film is placed at the upstream side and the polyethylene at the downstream side. As one would expect, the data shown in Figs. 3.2 and 3.4

highlight a marked difference between the water permeability coefficients of the studied polyamide bilayer systems.

The last element of the model proposed by Del Nobile et al. (2003b) is the water mass balance equation inside the package. In particular, the authors used the following assumptions to derive the water mass balance equation: (1) the water vapor present in the package headspace is always in equilibrium to the water absorbed into the packaged product and (2) the water vapor present in the package headspace behaves as an ideal gas. These two assumptions were experimentally tested in several works reported in the literature (Quast and Karel 1972; Bell and Labuza 2000). On the basis of the foregoing assumptions, the water mass equation can be written as follows:

$$\frac{dn_{Tot}(t)}{dt} = P_W^M\left(a_W^{est}, a_W^{in}(t)\right) \cdot S \cdot p_W^* \cdot \frac{a_W^{est} - a_W^{in}(t)}{\ell}, \tag{3.5}$$

where $n_{Tot}(t)$ is the total number of moles of water present inside the package at time t, $P_W^M\left(a_W^{est}, a_W^{in}(t)\right)$ is the water permeability coefficient of the bilayer film, $a_W^{in}(t)$ is the water activity inside the package at time t, a_W^{est} is the external water activity, and S is the surface area of the package. $P_W^M\left(a_W^{est}, a_W^{in}(t)\right)$ was evaluated using Eq. 3.2. $n_{Tot}(t)$ is the sum of two contributions: the moles of water present in the package headspace and the moles of water absorbed into the packed cereal-based dry product:

$$n_{Tot}(t) = n_{st}(t) + n_{ass}(t), \tag{3.6}$$

where $n_{st}(t)$ is the number of moles of water present in the package headspace at time t, and $n_{ass}(t)$ is the number of moles of water absorbed into the packaged product at time t. $n_{st}(t)$ depends on the water activity inside the package determined through the following expression:

$$n_{st}(t) = \frac{V_{st} \cdot p_W^* \cdot a_W^{in}(t)}{R \cdot T}, \tag{3.7}$$

where V_{st} is the volume of the package headspace. According to Eq. 3.1, the moles of water absorbed into the packaged food depend on the internal water activity through the following expression:

$$n_{ass}(t) = \frac{mp}{18} \cdot \frac{M_0 \cdot \Lambda \cdot \Gamma \cdot a_W^{in}(t)}{\left(1 - \Gamma \cdot a_W^{in}(t)\right) \cdot \left(1 - \Gamma \cdot a_W^{in}(t) + \Gamma \cdot \Lambda \cdot a_W^{in}(t)\right)}, \tag{3.8}$$

where mp is the mass of the packaged product. Substituting Eqs. 3.7 and 3.8 into Eq. 3.5 one obtains

$$\frac{d}{dt}\left[\frac{mp}{18}\cdot\frac{M_0\cdot\Lambda\cdot\Gamma\cdot a_W^{in}(t)}{\left(1-\Gamma\cdot a_W^{in}(t)\right)\cdot\left(1-\Gamma\cdot a_W^{in}(t)+\Gamma\cdot\Lambda\cdot a_W^{in}(t)\right)}+\frac{V_{st}\cdot p_W^*\cdot a_W^{in}(t)}{R\cdot T}\right]\qquad(3.9)$$

$$=P_W^M\left(a_W^{est},a_W^{in}(t)\right)\cdot S\cdot p_W^*\cdot\frac{a_W^{est}-a_W^{in}(t)}{\ell}.$$

It can be easily demonstrated that Eq. 3.9 can be rearranged in the following form:

$$\frac{da_W^{in}(t)}{dt}=\frac{P_W^M\left(a_W^{est},a_W^{in}(t)\right)\cdot S\cdot p_W^*\cdot\frac{a_W^{est}-a_W^{in}(t)}{\ell}}{\frac{V_{st}\cdot p_W^*}{R\cdot T}+\frac{mp\cdot M_0\cdot\Lambda\cdot\Gamma\cdot\left[1+\left(K\cdot a_W^{in}(t)\right)^2\cdot(\Lambda-1)\right]}{18\cdot\left[\left(1-\Gamma\cdot a_W^{in}(t)\right)\cdot\left(1-\Gamma\cdot a_W^{in}(t)+\Lambda\cdot\Gamma\cdot a_W^{in}(t)\right)\right]^2}}.\qquad(3.10)$$

Equation 3.10 is an ordinary differential equation where the unknown function is $a_W(t)$. It was numerically integrated by Del Nobile et al. (2003b) using a fourth-order Runge–Kutta formula (Press et al. 1989a). Once the function $a_W(t)$ is known, the amount of water absorbed into the packaged product during storage (i.e., the sole quality index of the packaged cereal-based dry product) can be easily determined using Eq. 3.1. As highlighted by the authors of this study, the major limitations to an extension of the proposed model to a general case is that an instantaneous equilibrium between the water vapor present in the package headspace and the water absorbed into the packaged product was hypothesized. Wherever this assumption is satisfied, the proposed model can be advantageously used to predict the SL of cereal-based dry foods or to properly design their package.

As an example, the model proposed by Del Nobile et al. (2003b) (i.e., Eq. 3.10) is used to simulate the storage behavior of two different dry foods packaged with the two hypothetical bilayer films. The characteristics of the package used to run the simulations of storage and packaging conditions are as follows: storage temperature 25 °C, water activity 0.70, initial water activity inside package 0.08, package surface 1,600 cm², package unfilled volume 500 cm³, product mass 500 g. As reported by these authors, the critical moisture content reported earlier corresponds to a water activity value of approximately 0.4, an intermediate value between 0.35 and 0.5, a range indicated as critical for many dry, crispy foods that become soft and texturally unacceptable (Labuza and Contreras-Medellin 1981). The SL of food/material combinations, as predicted by Del Nobile et al. (2003b) using Eq. 3.10, using the calculated fitting parameters ranged from 3.7 to 19.1 days, clearly underlining that the predicted SL strongly depends on the type of packaged food, whereas the type of bilayer used for the package has only a slight influence.

The model proposed by Del Nobile et al. (2003b) was also used to estimate the error made if the dependence of the water permeability coefficient on the water activity inside and outside the package was neglected. In particular, the authors also predicted the SL according to the following expression:

$$\frac{da_W^{in}(t)}{dt}=\frac{P_{W,Av.}^M\cdot S\cdot p_W^*\cdot\frac{a_W^{est}-a_W^{in}(t)}{\ell}}{\frac{V_{st}\cdot p_W^*}{R\cdot T}+\frac{mp\cdot M_0\cdot\Lambda\cdot\Gamma\cdot\left[1+\left(K\cdot a_W^{in}(t)\right)^2\cdot(\Lambda-1)\right]}{18\cdot\left[\left(1-\Gamma\cdot a_W^{in}(t)\right)\cdot\left(1-\Gamma\cdot a_W^{in}(t)+\Lambda\cdot\Gamma\cdot a_W^{in}(t)\right)\right]^2}},\qquad(3.11)$$

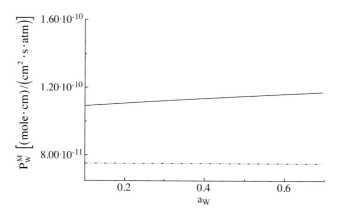

Fig. 3.6 Water permeability coefficients plotted as function of water activity at downstream side of polyamide film. *Horizontal line*: average water permeability coefficient obtained by averaging data reported in Fig. 3.5

where $P^M_{W,Av.}$ is the average value of the water permeability coefficient of the bilayer films in the interval $a_w = 0.1$, $a_w = 0.7$. $P^M_{W,Av.}$ was evaluated according to the following expression:

$$P^M_{W,Av.} = \frac{\int_{a_w=0.1}^{a_w=0.7} P^M_W\left(a^{est}_W, a^{in}_W(t)\right) da_W}{0.6}. \tag{3.12}$$

Equation 3.12 was numerically solved by Del Nobile et al. (2003b) using the extended Simpson's rule (Press et al. 1989b). A marked difference between data predicted using Eqs. 3.10 and 3.11 was found. In particular, the SL predicted by neglecting the dependence of the water permeability coefficient on the water activity inside and outside the package (Eq. 3.11) is approximately 90 % greater than the SL predicted using Eq. 3.10. To find the reasons for the observed large difference, Del Nobile et al. (2003b) plotted the water permeability coefficient of a bilayer film as a function of the water activity at the downstream side of the film (Fig. 3.6). The curve shown in the figure was predicted by means of Eq. 3.2, using the calculated fitting parameters. In particular, Del Nobile et al. (2003b) set the water activity at the upstream side of the film to 0.7 (i.e., the value of the water activity outside the package), whereas the water activity at the downstream side of the film is that reported on the abscissa. It is worth noting that data represent the water permeability coefficient of the investigated films in real working conditions. In Fig. 3.6 the value of $P^M_{W,Av.}$ for the investigated bilayer film is also reported. It is evident that there is a substantial difference between the average water permeability coefficient of the bilayer film and its water permeability coefficient in real working conditions. This difference is responsible for the marked discrepancy between the SL predicted by Eq. 3.10 and that predicted by Eq. 3.11. It is worth noting that if the permeability coefficient of the packaging film depends on the partial pressure of the permeating substance inside and outside the package, as was the case for the polymers investigated by Del Nobile et al. (2003b), then the permeation tests can give misleading information (Fig. 3.6). In these cases, a more accurate analysis of the permeation process, like that proposed earlier, is in order.

3.2.2 Potato Chips

Potato chip packaging systems are designed to keep the oxygen and water partial pressure in the package headspace as low as possible during storage. This is because the quality of this product depends on two quality indices: the extent of lipid oxidation and the amount of absorbed water (this latter is inversely related to the crispness of the product) (Quast et al. 1972; Quast and Karel 1972). To satisfy the preceding conditions, potato chips are currently packaged using nitrogen as the inert gas and polymeric films characterized by a low permeability to both oxygen and water vapor. As will be discussed in more details subsequently, an increase in the initial value of the oxygen partial pressure in the package headspace, increasing the lipid oxidation rate, leads to a decrease in the product SL. On the other hand, an increase in the initial value of water vapor partial pressure in the package headspace can yield either an increase or a decrease in the product SL, depending on the quality index responsible for its unacceptability. In fact, the SL increases if rancidity is responsible for the unacceptability of the product, whereas it decreases in the other case. Del Nobile (2001) provided an estimation of the potato chips' optimal headspace gas composition (in terms of nitrogen and water vapor) based on the use of a mechanistic model able to describe the product quality decay kinetic. In particular, in this study, the author first describes lipid oxidation and crispness as the main packaged-food quality indices, by means of empirical equations reported in the literature, then he combines the preceding information through the water and oxygen mass balance equation inside the package.

Lipid oxidation is a rather complex phenomenon with an evolution rate that depends on the oxygen partial pressure, water partial pressure, and extent of the lipid oxidation reaction. Quast and Karel (1972) proposed an equation to predict the lipid oxidation reaction rate of potato chips. The equation was also reported in a previous chapter, but for the sake of simplicity it is also reported below:

$$\frac{dExt(t)}{dt} = mp \cdot \left(Ext(t) + \frac{M_1 + M_2 \cdot Ext(t)}{\sqrt{\frac{p_W^{in}(t)}{p_W^*} \cdot 100}} \right) \cdot \left(\frac{p_{O_2}^{in}(t)}{M_3 + M_4 \cdot p_{O_2}^{in}(t)} \right), \quad (3.13)$$

where $Ext(t)$ is the extent of the oxidation reaction rate at time t, $p_W^{in}(t)$ is the package headspace vapor partial pressure at time t, $p_{O_2}^{in}(t)$ is the package headspace oxygen partial pressure at time t, mp is the mass of the packaged food, and the M_i are to be regarded as fitting parameters with no particular physical meaning. Crispiness is inversely related to the amount of water absorbed by the product. Quast and Karel (1972) also reported that the water sorption isotherm of potato chips can be described through Khun's isotherm:

$$c_W(t) = \frac{M_5}{100 \cdot \ln\left(\frac{p_W^{in}(t)}{p_W^*}\right)} + M_6, \tag{3.14}$$

where the M_i are fitting parameters.

Creating a mass balance on the oxygen and the water vapor contained in the package headspace, the following equations are obtained:

$$\frac{dp_{H_2O}^{in}(t)}{dt} = \frac{A \cdot P_W \cdot \frac{p_W^{out} - p_W^{in}(t)}{\ell}}{\frac{V_{st}}{R \cdot T} - \frac{M_5 \cdot mp}{M \cdot p_W^{in}(t) \cdot 100 \cdot \left[\ln\left(\frac{p_W^{in}(t)}{p_W^*}\right)\right]^2}}, \tag{3.15}$$

$$\frac{dp_{O_2}^{in}(t)}{dt} = \frac{R \cdot T}{V_{st}} \cdot \left(A \cdot P_{O_2} \cdot \frac{p_{O_2}^{out} - p_{O_2}^{in}(t)}{\ell}\right)$$
$$+ \frac{R \cdot T}{V_{st}} \cdot \left[\frac{mp \cdot (44.615 \cdot 10^{-9})}{3,600} \cdot (Ext(t).\right.$$
$$\left. + \frac{M_1 + M_2 \cdot Ext(t)}{\sqrt{\frac{p_W^{in}(t)}{p_W^*}} \cdot 100} \cdot \left(\frac{p_{O_2}^{in}(t)}{M_3 + M_4 \cdot p_{O_2}^{in}(t)}\right)\right], \tag{3.16}$$

where p_W^{out} is the water vapor partial pressure outside the package, $p_{O_2}^{out}$ is the oxygen partial pressure outside the package, and P_W and P_{O_2} are the water vapor and oxygen permeability of the packaging film. Equation 3.15 was derived assuming that (1) the water vapor present in the package headspace is always in equilibrium with the water absorbed into the packaged product and (2) both the water vapor and oxygen present in the package headspace behaves as an ideal gas.

Equations 3.13, 3.15, and 3.16 are a set of three ordinary differential equations with three unknown variables [i.e., the two packaged-food quality indices: Ext(t), $p_W^{in}(t)$, and $p_{O_2}^{in}(t)$]. Once the initial packaging conditions are known, the differential equations can be numerically solved (Press et al. 1989a). In this way, it is possible to predict the kinetic decay of the two packaged-food quality indices and, consequently, the product SL.

This model was used by Del Nobile (2001) to simulate potato chip quality loss during storage. The author calculated the potato chip SL on the basis of the simulated quality decay kinetic curve. Moreover, the influence of the initial water

vapor partial pressure in the package headspace and the barrier properties of the
packaging film on the SL of the packaged product were also addressed. In particular, the storage simulations were made for a bag-type package with the following
characteristics: the area of the package exposed to the mass flux was equal to
400 cm^2, the headspace volume was equal to 1,435 cm^3, and the weight of the
product was equal to 65 g. The study conducted by Del Nobile (2001) was carried
out using a multilayer film commercially used for packaging potato chips, obtained
by laminating two films of PE coated with polyvinylidene chloride (PVdC). The
author assumed that the water and oxygen permeability coefficient of the aforementioned bilayer flexible film did not depend on the water vapor partial pressure at
the film sides and calculated the water and oxygen permeability coefficient of the
packaging film using Eq. 2.29. The single-layer permeability coefficients were
taken from the literature (Myers et al. 1961; Rogers et al. 1956; Yasuda and Stannett
1975). The storage and packaging conditions considered for the simulations were as
follows: storage temperature 37 °C, oxygen partial pressure outside the package
0.21 atm, water vapor partial pressure outside the package 24.8×10^{-3} atm
(i.e., 40 % relative humidity), gas flushed into the package composed by humidified
nitrogen with a humidity ranging between 0.1 % and 32 %, and initial value of
Ext(t) of zero.

Since the quality of potato chips depends on the amount of absorbed water and
on the extent of lipid oxidation reaction, Del Nobile (2001) proposed expressing
the total quality of the product through two normalized quality indices defined
as follows:

$$Q_W(t) = 1 - \frac{p_W^{in}(t)}{p_W^{max}}, \tag{3.17}$$

$$Q_O(t) = 1 - \frac{Ext(t)}{Ext^{max}}, \tag{3.18}$$

where $Q_W(t)$ is the normalized quality index related to the absorbed water at time t,
$Q_O(t)$ is the normalized quality index related to the extent of lipid oxidation at
time t, p_W^{max} is the threshold value for $p_{H_2O}^{in}(t)$, and Ext^{max} is the threshold value for
Ext(t). As reflected by literature data, the value of p_W^{max} is equal to $20 \cdot 10^{-3}$ atm
(Quast and Karel 1972), whereas that of Ext^{max} is equal to 1,200 $\frac{\mu lO_2(STP)}{g}$ (Quast
and Karel 1972). The product becomes unacceptable whenever either one of the
two quality indices exceeds its threshold value or one of the two normalized quality
indices reaches zero.

In the hypothetical case where the packaged product quality is related only to
$Q_W(t)$, the SL of the product would decrease with the initial value of the water vapor
partial pressure in the package headspace ($p_W^{initial}$). On the other hand, if the
packaged product quality were related only to $Q_O(t)$, then the SL of the product
would increase with $p_W^{initial}$. The curve representing the "true" SL versus $p_W^{initial}$ was

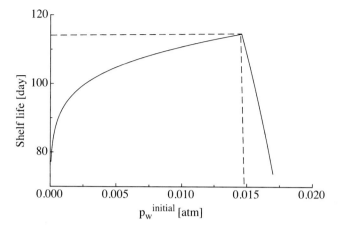

Fig. 3.7 Shelf life as function of $p_w^{initial}$. PE layers were 40 μm thick; PVdC was 4 μm thick

obtained, by definition, by superimposing the two trends, as shown in Fig. 3.7. As can be inferred, the curve presents a maximum (114.5 day) at $p_W^{initial}$ equal to 0.0146 atm, suggesting that $p_W^{initial}$ can be used as a packaging design variable to prolong the SL of potato chips. In fact, according to the data reported in Fig. 3.7, and considering that nitrogen used to package potato chips has only small traces of water vapor, substituting nitrogen with a mixture of water vapor and nitrogen, characterized by a value of $p_W^{initial}$ equal to 0.0146 atm, yields an increase in the product SL. The value of $p_W^{initial}$ corresponding to the maximum of the curve is the optimal headspace water vapor partial pressure $\left(\left[p_W^{initial} \right]_{opt} \right)$. Its value depends on the storage conditions and on the permeability and selectivity to the water vapor and oxygen of the packaging film. To evaluate the influence of the film barrier properties on $\left[p_W^{initial} \right]_{opt}$, Del Nobile (2001) calculated its value changing the PVdC thickness, ℓ_{PVdC}, from 1 to 6 μm. Increasing the value of ℓ_{PVdC}, the film permeability decreases, while its mechanical properties and perm-selectivity to oxygen and water vapor are substantially unmodified. In particular, the aforementioned authors calculated the SL of the product when $p_W^{initial}$ was equal to $62 \cdot 10^{-6}$ atm (percentage relative humidity equal to 0.1 %, i.e., technical nitrogen) as a function of ℓ_{PVdC}; the optimal shelf life (OSL) of the product when $p_W^{initial}$ was equal to $\left[p_W^{initial} \right]_{opt}$ as a function of ℓ_{PVdC}; the percentage difference between the SL and OSL (ΔSL%), calculated as $\Delta SL\% = \frac{OSL - SL}{SL} \cdot 100$ as a function of ℓ_{PVdC}. It was noted that OSL was always higher than SL. Moreover, a value of ΔSL% as high as 48 % was obtained for a value of ℓ_{PVdC} equal to 4.4×10^{-4} cm. In addition, contrary to what one would expect, the curve representing SL versus ℓ_{PVdC} showed a minimum. In fact, increasing the value of ℓ_{PVdC} the water vapor and oxygen permeability reduced by about the same amount. This reduced the rate at which both substances penetrated into the package and, consequently, decreased the rate at which their respective partial pressures increased inside the package during storage.

As previously reported, a reduction in the water vapor and oxygen partial pressure can cause either an increase or a decrease in the lipid oxidation rate, as is evident from Eq. 3.13. In fact, it is not possible, on the basis of simple intuition, to predict the effect of an increase in ℓ_{PVdC} on the lipid oxidation rate. According to simulations by Del Nobile (2001) for values of ℓ_{PVdC} lower than 2.5 µm, the effect related to water vapor was dominant, whereas for higher values of ℓ_{PVdC} the contrary was true.

3.3 Unsteady-State Conditions

There are certain circumstances where the mass exchanged during a transient state cannot be neglected for SL modeling. In these cases the mass flux between the inside and outside of the package cannot be calculated using Eq. 1.4. To introduce a new approach to calculating the mass exchanged during a transient state, two distinct cases will be presented in what follows: the SL prediction of a carbonated beverage stored under fluctuating temperature conditions and the SL prediction of extra virgin olive oil. However, before illustrating these specific cases, the need to change the way in which the mass exchanged between the inside and the outside of a package is calculated will be addressed. To this end, the simple case of carbonated beverages stored under constant temperature conditions will be considered.

The quality of a carbonated beverage is simply related to the carbon dioxide concentration inside the bottle. The carbon dioxide molecules inside the bottle are in turn partially dissolved in the liquid phase and partially dispersed in the bottle headspace. It is generally assumed that the amount of carbon dioxide dissolved in the liquid phase is proportional to the headspace carbon dioxide partial pressure, i.e., Henry's law is obeyed:

$$n_{CO_2}^{D} = H \cdot p_{CO_2}^{ins},$$ (3.19)

where $n_{CO_2}^{D}$ is the number of carbon dioxide moles dissolved in the liquid phase at equilibrium, $p_{CO_2}^{ins}$ is the carbon dioxide partial pressure in the bottle headspace, and H is the carbon dioxide partition coefficient.

Due to the difference between the carbon dioxide partial pressure inside and outside the bottle, carbon dioxide permeates through the bottle wall. As a first-order approximation it can be assumed that the amount exchanged can be calculated using the same equation used for flexible packaging (i.e., Eq. 4 of part I), which implicitly neglects the transient state:

$$J_{SS}^{CO_2} = P_{CO_2} \cdot \frac{p_{CO_2}^{ins} - p_{CO_2}^{ext}}{\ell},$$ (3.20)

where $J_{SS}^{CO_2}$ is the steady-state carbon dioxide mass flux, $p_{CO_2}^{ext}$ is the external carbon dioxide partial pressure, and P_{CO_2} is the carbon dioxide permeability coefficient.

It can further be assumed that the external carbon dioxide partial pressure is equal to zero:

$$J_{SS}^{CO_2} = P_{CO_2} \cdot \frac{p_{CO_2}^{ins}}{\ell}.$$ (3.21)

Formulating the carbon dioxide mass balance equation inside the bottle in the same way it was done earlier for flexible packaging films, assuming that pseudo-steady-state conditions hold, the following expression is obtained:

$$\frac{dn_{CO_2}(t)}{dt} = -\frac{A \cdot P_{CO_2}}{\ell} \cdot p_{CO_2}^{ins}(t),$$ (3.22)

where $n_{CO_2}(t)$ is the total number of carbon dioxide moles inside the bottle at time t. The left-hand side of Eq. 3.22 is the accumulation term, whereas the right-hand side is the rate at which carbon dioxide is lost from the inside of the bottle. Since the moles of carbon dioxide inside the bottle are partially dissolved in the liquid phase and partially dispersed in the bottle headspace, it can be written as

$$n_{CO_2}(t) = n_{CO_2}^{HS}(t) + n_{CO_2}^{D}(t),$$ (3.23)

where $n_{CO_2}^{HS}(t)$ is the number of carbon dioxide moles in the bottle headspace at time t, and $n_{CO_2}^{D}(t)$ is the number of carbon dioxide moles dissolved in the liquid phase at time t. It is generally assumed that carbon dioxide in a bottle's headspace behaves as an ideal gas; if this is true, then the following expression is obtained for the moles of carbon dioxide in the bottle headspace:

$$n_{CO_2}^{HS}(t) = \frac{V_{st} \cdot p_{CO_2}^{ins}(t)}{R \cdot T}.$$ (3.24)

It can be further hypothesized that the carbon dioxide in the bottle headspace is always in equilibrium with gas in the liquid phase. In this case the dissolved moles are directly proportional to the gas headspace partial pressure:

$$n_{CO_2}^{D}(t) = H \cdot p_{CO_2}^{ins}(t).$$ (3.25)

Substituting Eqs. 3.23, 3.24, and 3.25 into Eq. 3.22 and rearranging, the following expression is obtained:

$$\frac{dp_{CO_2}^{ins}(t)}{dt} = -\frac{\frac{A \cdot P_{CO_2}}{\ell}}{\frac{V_{st}}{R \cdot T} + H} \cdot p_{CO_2}^{ins}(t).$$ (3.26)

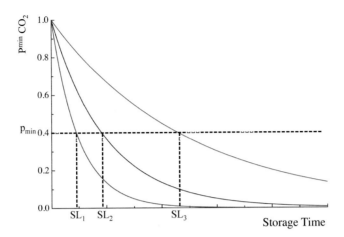

Fig. 3.8 Shelf life defined as intercept of quality decay *curves* with *horizontal line* at $p_{CO_2}^{min}$

Equation 3.26 is an ordinary differential equation whose unknown function is the headspace carbon dioxide partial pressure (i.e., the only packaged-food quality index). The solution of Eq. 3.26 is as follows:

$$p_{CO_2}^{ins}(t) = p_{CO_2}^{ins}(0) \cdot \exp\left[-\frac{A \cdot P_{CO_2}}{\ell \cdot \left(\frac{V_{sl}}{R \cdot T} + H\right)} \cdot t\right], \qquad (3.27)$$

where $p_{CO_2}^{ins}(0)$ is the initial carbon dioxide partial pressure in the bottle headspace. By plotting $p_{CO_2}^{ins}(t)$ as a function of storage time as predicted by Eq. 3.27, the carbon dioxide depletion rate decreases as the bottle wall thickness increases. To determine the carbonated beverage SL, a threshold value for the packaged-food quality index [i.e., $p_{CO_2}^{ins}(t)$] must be generally set. For the sake of simplicity, it was assumed that the aforementioned threshold was equal to $p_{CO_2}^{min}$. The SL can be easily found from the intercepts of the quality decay curves reported in Fig. 3.8 with the horizontal line at $p_{CO_2}^{min}$. Alternatively, this is the expression that relates carbonated beverage SL to bottle thickness:

$$SL = \left\{\ln\left[\frac{p_{CO_2}^{min}}{p_{CO_2}^{ins}(0)}\right] \cdot \frac{\left(\frac{V_{sl}}{R \cdot T} + H\right)}{A \cdot P_{CO_2}}\right\} \cdot \ell. \qquad (3.28)$$

Because the term $\left\{\ln\left[\frac{p_{CO_2}^{min}}{p_{CO_2}^{ins}(0)}\right] \cdot \frac{\left(\frac{V_{sl}}{R \cdot T} + H\right)}{A \cdot P_{CO_2}}\right\}$ is a constant, Eq. 3.28 can be rewritten as follows:

$$SL = \Theta \cdot \ell, \qquad (3.29)$$

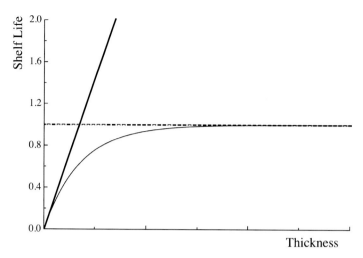

Fig. 3.9 Carbonated beverage SL increases with bottle thickness. According to Eq. 3.29, SL would linearly increase with bottle thickness; real carbonated beverage SL first increases with thickness, then reaches plateau

where $\Theta = \left\{ \ln\left[\dfrac{p_{CO_2}^{min}}{p_{CO_2}^{ins}(0)} \right] \cdot \dfrac{\left(\frac{V_{st}}{R \cdot T}+H\right)}{A \cdot P_{CO_2}} \right\}$.

A hypothetical test, in which it is possible to increase the bottle thickness as much as desired, is now taken into account. According to Eq. 3.29 the carbonated beverage SL would linearly increase with bottle thickness (Fig. 3.9). In fact, the real carbonated beverage SL first increases with thickness and then reaches a plateau. The fact that there is a difference between what is predicted by Eq. 3.29, which was derived assuming pseudo-steady-state conditions, and the reality is due to the fact that the pseudo-steady-state-conditions hypothesis does not hold for thicker package structures. In fact, neglecting the transient state, the amount of permeant, such as carbon dioxide in the case illustrated earlier, dissolved in the package structure wall is neglected. In conclusion, it is generally reasonable to neglect the amount of permeant dissolved inside a package wall for thinner packages, such as flexible films, but not for thicker packages, such as plastic bottles.

3.3.1   Carbonated Beverages: Fluctuating Temperature Conditions

As highlighted earlier, there are many cases where the SL of a packaged product is strictly related to the transport of small molecules provided by the film or the container wall. A typical example of this situation is represented by soft drinks and beer bottled in polymeric containers. The carbonation level of these beverages decreases with time due to gas sorption into the bottle wall and to the subsequent diffusion through it. Their sensory quality is strongly affected by their gas content and in most cases a decrease in the carbon dioxide content as low as 10 % causes their taste to become flat and, hence, unacceptable to the consumer (Fenelon 1973).

In principle, the prediction of the SL of carbonated beverages bottled in a plastic container is easy. It can be obtained by simply modeling the dynamics of carbon dioxide leaving the container. There is a considerable practical interest in solving this problem because it represents the only rational way to design a plastic bottle, in particular its thickness. However, to reduce the difficulty of the calculations, the modeling procedure is often oversimplified and gives unrealistic results (Masi and Paul 1982). To theoretically estimate gas loss from inside pressurized plastic containers, it is necessary to solve the differential mass balance equation under appropriate boundary and initial conditions. Three issues are relevant in this connection. Concerning the first issue, as was reported earlier, the use of simple pseudo-steady-state conditions is not adequate in quantifying the mass exchanged by thicker containers such as plastic bottles (Barrer et al. 1958; Michaels et al. 1963; Vieth et al. 1966; Paul 1979). Regarding the second issue, it has been demonstrated that the mass transport model used to describe the barrier properties of bottle materials has a great influence on SL prediction of carbonated beverages. The simple approach, which is usually used in practical applications, assumes that Henry's law describes gas partitioning between the gas phase and the polymer and that mass transfer inside the bottle wall is governed by Fick's model. This gives rise to the underestimation of the barrier properties of the materials generally used for this application, and consequently the predicted SL of carbonated beverages is much shorter than in reality (Masi and Paul 1982). The third issue concerns the temperature conditions used to simulate bottle storage. In fact, in predicting the SL of biologically stable products, their thermal history is generally not taken into account. In many cases, however, this approximation is not appropriate, and the predicted SL can be quite different from the real one. Commonly, carbonated beverages are bottled at low temperatures in order to favor gas retention during this operation. Once the bottle is capped, no specific precautions are taken during storage and distribution. Because of the large storage volumes required, distributors and grocery stores often store pallets of bottles of carbonated beverages outdoors without paying any attention to temperature rises due to sunlight exposure. Since most carbonated beverages are stable from a biological point of view, it is rare for the thermal history experienced by the bottled beverage during storage and distribution to be taken into account in design applications. However, this approximation may lead to wrong conclusions.

The study conducted by Del Nobile et al. (1997) was aimed at showing the error that is possible by underestimating the importance of the bottle thermal history. In particular, three cases were considered by the authors: (1) the SL of the bottled beverage was estimated assuming that the storage temperature was constant and equal to room temperature for the entire storage period; (2) the temperature of the bottle varied during the storage period, but for the sake of simplicity, in performing the calculation the temperature was kept constant and equal to the average temperature of the storage period; (3) the temperature of the bottle of the carbonated

Fig. 3.10 Mechanism for
CO_2 loss in typical 2L plastic
soft-drink bottle

Inside$\rightleftharpoons$Outside

beverage varied during the storage period and during the day was examined. In what follows, these three cases will be illustrated and discussed in detail.

For the simulation Del Nobile et al. (1997) used a typical 2-L plastic soft-drink bottle made up of PET. To facilitate mathematical analysis, the authors made the following assumptions: (1) the bottle was considered an equivalent cylinder of constant thickness, ℓ, with approximately the same total capacity and exposed area as a real bottle; (2) the carbon dioxide mass diffusion through the bottle wall is monodimensional and occurs only in the radial direction; (3) such a container represents a closed system where carbon dioxide dissolved in the liquid phase, whose volume is V_L, is supposed to be at all times in equilibrium with the carbon dioxide in the gas phase, whose volume is V_G; (4) the driving force for carbon dioxide permeation is assumed to be uniform over the container's internal surface and, at all times, equal to the partial pressure of carbon dioxide in the container headspace (the partial pressure of carbon dioxide in the external atmosphere was set equal to zero), which is a decreasing function of storage time; (5) the equilibrium existing between the phases can be described by Henry's law:

$$C_{CO_2} = H \cdot p_{CO_2}^{ins}, \tag{3.30}$$

where C_{CO_2} is the carbon dioxide molar concentration in the liquid phase.

The mechanism proposed by Del Nobile et al. (1997) for gas loss is shown in Fig. 3.10. The gas loss from the interior of the bottle was evaluated via the solution of the nonlinear parabolic differential equation derived from the gas mass balance made inside the bottle wall. Accordingly, the concentration profile $C_{CO_2}(t, x)$ was evaluated at each time and from it the total flux of gas leaving the inside of the bottle was estimated through Fick's first law with a diffusion coefficient dependent on the local carbon dioxide concentration. To write the gas mass balance equation, the solubilization and diffusion processes through PET, which is a glassy polymer at room temperature, must be addressed first. As reported in the literature, many investigators have demonstrated that the most appropriate way to describe gas transport through glassy polymers is the so-called dual sorption–dual mobility model (Vieth et al. 1966; Koros 1980; Chern et al. 1983; Koros and Hellums 1990).

The dual sorption–dual mobility model is based on the assumption that a glassy polymer can be envisaged as a polymer matrix at equilibrium in which frozen microvoids are uniformly dispersed. The presence of frozen microvoids is used to take into account the state of being out of equilibrium of a glassy polymer. In fact, the volume of frozen microvoids accounts for the excess free volume present in a glassy polymer. As a consequence, the total gas concentration absorbed into a glassy polymer, such as PET at room temperature, is the result of two independent contributions. One mode is related to the gas dissolution into the polymeric matrix at equilibrium, and it corresponds to the expectation for liquids and rubbery polymers, that is, that the gas molecules are randomly dispersed in the polymeric matrix at equilibrium. Gas dissolved according to the mechanism described earlier follows Henry's law. The additional mode is related to gas adsorption onto the surface of the frozen microvoids; the hypothesized adsorption mechanism is analogous to that generally used to describe gas and vapor adsorption in zeolites. It is generally expressed through the Langmuir adsorption isotherm. Therefore, the relationship between the carbon dioxide adsorbed at equilibrium in a glassy polymer and its partial pressure is

$$C_{CO_2}^{Tot} = C_{CO_2}^{D} + C_{CO_2}^{H} = k_D \cdot p_{CO_2} + \frac{C_H' \cdot b \cdot p_{CO_2}}{1 + b \cdot p_{CO_2}}, \qquad (3.31)$$

where $C_{CO_2}^{Tot}$ is the carbon dioxide adsorbed at equilibrium in the polymeric matrix, the superscripts H and D identify the two different populations of molecules, one that is adsorbed into holes and another that is dissolved into the polymer matrix, respectively, k_D is the partition coefficient relative to the gas population dissolved according to Henry's law, b is a parameter that describes the gas–polymer affinity, and C_H' is the adsorption capacity of the frozen microvoids.

According to the dual sorption–dual mobility model, the two populations of permeating molecules have different mobilities and are always in local equilibrium with each other. The modified version of Fick's first law for gas diffusion in glassy polymers is

$$J_{CO_2} = -D_{CO_2}^{D} \cdot \frac{\partial C_{CO_2}^{D}}{\partial x} - D_{CO_2}^{H} \cdot \frac{\partial C_{CO_2}^{H}}{\partial x}, \qquad (3.32)$$

where J_{CO_2} is the total carbon dioxide mass flux, $D_{CO_2}^{D}$ and $D_{CO_2}^{H}$ are the diffusivities of the dissolved and adsorbed molecules, respectively, and x is the spatial coordinate.

The assumption of local equilibrium between the two adsorbed populations allows to write Eq. 3.32 in a more useful way:

$$J_{CO_2} = -D_{eff} \cdot \frac{\partial C_{CO_2}^{Tot}}{\partial x}, \qquad (3.33)$$

where D_{eff} is the effective diffusion coefficient; it is a concentration-dependent coefficient given by the following expression:

$$D_{eff} = D_{CO_2}^D \cdot \left[\frac{1 + \dfrac{\left(\dfrac{D_{CO_2}^H}{D_{CO_2}^D}\right) \cdot \left(\dfrac{c_H' \cdot b}{k_D}\right)}{\left(1 + b \cdot p_{CO_2}\right)^2}}{1 + \dfrac{\dfrac{c_H' \cdot b}{k_D}}{\left(1 + b \cdot p_{CO_2}\right)^2}} \right].$$ (3.34)

The equation describing the differential mass balance on CO_2 made inside the bottle wall has the following form:

$$\frac{\partial C_{CO_2}^{Tot}(t, x)}{\partial t} = \frac{\partial}{\partial x}\left(D_{eff} \cdot \frac{\partial C_{CO_2}^{Tot}(t, x)}{\partial x}\right).$$ (3.35)

According to the scheme proposed by Del Nobile et al. (1997) to describe carbon dioxide loss from plastic bottles, the appropriate boundary and initial conditions that apply to the preceding carbon dioxide mass balance equation are as follows:

$$\begin{cases} t = 0, \quad 0 < x < 1 \quad \Rightarrow \quad p_{CO_2}(t) = 0, \\[2mm] t \geq 0, \quad x = 1 \quad \Rightarrow \quad p_{CO_2}(t) = 0, \\[2mm] t \geq 0, \quad x = 0 \quad \Rightarrow \quad p_{CO_2}(t) = \dfrac{n_{CO_2}^{Tot}(t)}{\dfrac{V_G}{R \cdot T} + H \cdot V_L}, \end{cases}$$ (3.36)

where $n_{CO_2}^{Tot}(t)$ is the total number of moles inside the container given by the sum of the moles of CO_2 in the bottle headspace (i.e., the gas phase) and the moles of CO_2 dissolved in the liquid phase.

Del Nobile et al. (1997) numerically solved Eq. 3.34 with appropriate initial and boundary conditions (Eq. 3.36) using an implicit finite-difference discretization scheme. The package conditions as well as bottle characteristics are as follows: temperature 5 °C, initial headspace pressure 4 atm, gas phase CO_2, liquid phase carbonated water, container made up of PET, container thickness 0.04 cm, surface area of container 55 cm², volume container 119 cm³, surface area of liquid phase 931.7 cm², liquid phase volume 2,000 cm³. Del Nobile et al. (1997) calculated $n_{CO_2}^{Tot}(t)$ by subtracting from the initial number of moles the amount of carbon dioxide that left the interior of the container. In the case where the simulation was carried out by considering that the temperature varied continuously at any step, the parameters needed to describe gas sorption and transport into the polymer wall were updated by considering their dependence on temperature.

The material parameters used by the authors were taken from the literature (Koros and Paul 1978). To facilitate the updating of the parameters in the computer routine, it was necessary to represent in a mathematical form the dependence of the transport and sorption parameters relative to water–CO_2 and PET–CO_2 systems on temperature. For the sake of simplicity, the mathematical equations used to predict the aforementioned parameters are reported as follows:

$$\begin{cases} H = 4.9976 \cdot 10^{-9} \cdot \exp\left(\dfrac{2626.1}{T}\right), \\[2mm] k_D = 7.8354 \cdot 10^{-3} \cdot \exp\left(\dfrac{1134.3}{T}\right), \\[2mm] C_H' = 13.252 - \dfrac{1.7951 \cdot 10^4}{T} + \dfrac{4.8595 \cdot 10^6}{T^2}, \\[2mm] b = 2.4638 \cdot 10^{-3} \cdot \exp\left(\dfrac{1495.6}{T}\right), \\[2mm] D_{CO_2}^D = 6.5148 \cdot 10^{-3} \cdot \exp\left(\dfrac{-4470.5}{T}\right), \\[2mm] D_{CO_2}^H = 8.2863 \cdot 10^{-3} \cdot \exp\left(\dfrac{-5406.5}{T}\right). \end{cases} \qquad (3.37)$$

To calculate the SL of the carbonated beverage, Del Nobile et al. (1997) assumed that the soft drink was capped at 5 °C. It was hypothesized that carbon dioxide was the only gas present in the bottle headspace, and immediately after capping its pressure was equal to 4 atm. It was also assumed that the bottle was opened for consumption at a temperature of 5 °C, while the temperature during distribution and storage was variable. The headspace carbon dioxide pressure predicted by the authors, $p_{CO_2}(t, \ T)$, is a function of time and temperature. Since simulations were performed at several temperatures, all pressure decay data were rescaled properly. Accordingly, the number of moles of carbon dioxide inside the container was converted to the equivalent carbon dioxide pressure at 5 °C.

To reproduce the thermal history experienced by a bottle during transportation and outdoor storage without any protection, Del Nobile et al. (1997) assumed that the daily average temperature of the soft drink changed during the year according to a sinusoidal function having an average value equal to 17.5 °C, a fluctuation amplitude of ±12.5 °C, and a period of 365 days. It was further assumed that the temperature varied during the day around the average temperature according to an asymmetric sine function with a period of 24 h and amplitude of 25 °C. The asymmetry was introduced to account for thermal spikes due to exposure to sunlight. Figure 3.11 shows the thermal history of the soft-drink bottle as described by the following equation:

$$T = 17.5 + 12.5 \cdot \sin\left(\frac{2 \cdot \pi \cdot t}{31536000}\right) + 12.5 \cdot \sin\left(\frac{2 \cdot \pi \cdot t}{86400}\right) + 7.5 \cdot \sin\left(\frac{2 \cdot \pi \cdot t}{86400}\right).$$

$$(3.38)$$

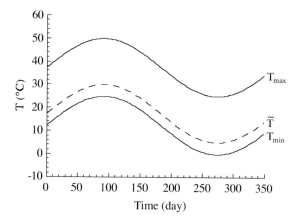

Fig. 3.11 Thermal history of soft drink bottle as described by Eq. 3.38

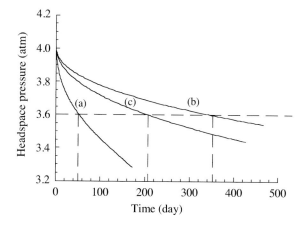

Fig. 3.12 Pressure decay inside plastic bottle in three different cases: (**a**) temperature was assumed to fluctuate according to Eq. 3.38; (**b**) temperature of carbonated beverage was considered to be constant during entire storage period and set equal to 20 °C; (**c**) temperature varied according to Eq. 3.38; however, for sake of simplicity it was averaged over a period of 70 days, obtaining a value equal to 28.9 °C

Del Nobile et al. (1997) considered two other cases in addition to the one where the storage temperature was assumed to fluctuate according to Eq. 3.38. In the first case, the temperature of the carbonated beverage was considered constant during the entire storage period and was set equal to 20°C; in the other case, the temperature varied according to Eq. 3.38; however, for the sake of simplicity the temperature was averaged over a period of 70 days, obtaining a value equal to 28.9 °C. This value was then used in the calculation that was carried out with a constant temperature. Figure 3.12 shows the pressure decay inside a plastic bottle as predicted by Del Nobile et al. (1997) in the three cases mentioned earlier. As can be inferred from the data, there is a remarkable difference in the SL relative to the three situations. This confirms the concern about the adequacy of some assumptions regarding the storage temperature that are generally made in designing carbonated-beverage containers. In particular, if one assumes that the temperature of the soft drink remains constant and equal to room temperature for the entire period of time between bottling and consumption, the simulation made by this study predicts a SL value longer than 1 year. Therefore, on this basis one would conclude that PET bottles commonly used for this application could provide an adequate protection to the carbonated beverage over a period of time that is more than enough for a normal distribution cycle. One would arrive at a similar conclusion under the assumption that the temperature is constant and equal to the average temperature of the distribution and storage period. The temperature used to perform the calculations in the second case was 9 °C higher than the temperature used in the first case, and the SL of the carbonated beverage decreases from 352 to 206 days. Although the SL reduction is quite significant, it may not be considered relevant from a practical point of view. In fact, a SL of 7 months is still more than adequate for commercial purposes. However, both results are based on approximations that are false, and the predicted SL is overestimated. In the first case, the fact that during storage and distribution the temperature of soft drinks can often fluctuate in a significant manner is not taken into proper account. Secondly, the influence of temperature on the sorption and transport parameters is underestimated. In fact, by averaging the temperature and using in the calculations the corresponding parameters one implicitly assumes that the transport and sorption parameters change linearly with temperature, but this is far from reality. When the thermal history of the bottle is considered and the proper relationships that describe the dependence of the physical parameters on the temperature are used, one finds that under conditions comparable to those occurring during distribution and outdoor storage, the SL of a carbonated beverage is less than 2 months, which is remarkably smaller than that predicted in the other two cases. Moreover, the SL predicted by Del Nobile et al. (1997) in this last case is comparable to the time that is usually required to distribute and sell soft drinks. Therefore, the results obtained by these authors suggest that there is a real risk that many bottles of soft drinks may reach the consumer at the very end of their SL.

3.3.2 Virgin Olive Oil

The SL of bottled vegetable oil is limited by the auto-oxidation of unsaturated fatty acids with the formation of hydroperoxides. The decomposition of hydroperoxides gives rise to different compounds, some of which are volatile and responsible for the sensory degradation of the oil (Frankel 1998). Among vegetable oils, virgin olive oil (VOO) shows the highest oxidation stability due to its low content of polyunsaturated fatty acids and the presence of natural antioxidants, such as tocopherols and, mainly, phenol compounds (Papadopulos and Boskou 1991; Tsimidou et al. 1992; Blekas et al. 1995; Satue et al. 1995).

Glass containers are generally preferred to plastic for bottling virgin olive oil. This is due in part to marketing aspects and in part to the better performance of glass containers compared to their plastic counterparts. Del Nobile et al. (2003a) presented a work to demonstrate the limits, in terms of performance, related to the use of plastic containers and to evaluate the efficiency of innovative containers in prolonging the SL of VOO by simulating the behavior of bottled oil using a mechanistic approach. In particular, in this work the influence of both oxygen diffusivity and wall thickness of the plastic container on the quality decay kinetic of bottled VOO was addressed. The innovative packaging solutions tested by Del Nobile et al. (2003a) were based on the use of oxygen scavengers that, by reacting with oxygen, reduce the rate at which oxygen permeates through the bottle wall. In fact, it has often been reported in the literature (Vermeiren et al. 1999) that oxygen scavengers can be successfully used to prolong the SL of foods whose quality decay kinetic depends on the oxygen concentration inside the package. The behavior of two bottles containing oxygen scavengers was simulated: a plastic bottle in which the oxygen scavenger is uniformly dispersed into the bottle wall and a glass bottle internally coated with a polymer in which the oxygen scavenger is uniformly dispersed. For comparative purposes Del Nobile et al. (2003a) also assessed the effect of a reduction in the amount of oxygen dissolved in the oil prior to bottling on hydroperoxide evolution during storage. The mathematical model used by the authors to determine the effectiveness of the containers in prolonging the VOO SL was derived assuming the average hydroperoxide concentration as a measure of the oil quality. It was obtained by combining the mass balance equations of oxygen and hydroperoxides with that describing the rate of hydroperoxide formation and decomposition. To validate the model, the authors monitored the average hydroperoxide concentration of the product bottled in glass, PET, and in an experimental polymeric container composed of a starch/polycaprolactone (PCL) blend stored at 40 °C. In what follows, the mechanistic model proposed by Del Nobile et al. (2003a) is presented in detail.

During storage of bottled virgin olive oil hydroperoxides are formed through the oxidation of unsaturated fatty acids and consumed by hydroperoxide breakdown reactions. In the first stage of oxidation, when the oxygen concentration is close to saturation, the rate at which hydroperoxides are consumed is lower than the rate at which they are produced through the auto-oxidation of unsaturated fatty acids, leading to an increase in hydroperoxide concentration. As the lipid oxidation reaction proceeds, oxygen is consumed to form hydroperoxides. This causes, first, the formation of an oxygen concentration gradient in the bottled oil, which in turn brings about the permeation of external oxygen through the wall of the plastic container, and, second, an increase in the rate at which the hydroperoxides break down. As a result of these phenomena, concentrations of both local oxygen and hydroperoxides decrease. Given the foregoing scenario during oil storage, to properly describe the quality decay kinetic of bottled virgin olive oil, it is therefore necessary to develop a mathematical model that can predict the time course of oxygen and hydroperoxide concentrations in the bottled oil during storage. To this end, the authors of the work assumed that the bottle could be represented by a cylinder composed of an outer shell made of plastic or glass and of an internal oil core, that oxygen mass diffusion was monodimensional and occurred only in the radial direction, and that the diffusive mass flux of hydroperoxides through both the olive oil and the container wall was considered negligible.

Under the foregoing restrictions the mass balance equation of the hydroperoxides dissolved in bottled oil is as follows:

$$\frac{\partial C_{ROOH}}{\partial t} = R_F - R_D, \tag{3.39}$$

where C_{ROOH} is the local hydroperoxide concentration in the bottled oil, R_F is the rate at which hydroperoxides are formed, and R_D is the rate at which they are decomposed.

Several models have been reported in the literature to describe the rate at which hydroperoxides are formed through the oxidation of unsaturated fatty acids (Quast and Karel 1972; Quast et al. 1972); the model adopted by Del Nobile et al. (2003a) to design a plastic oil bottle was derived from that proposed by Quast et al. (1972) to describe lipid oxidation in potato chips. Assuming that the relative humidity is constant during storage, the model proposed by Quast et al. (1972) is further simplified to the following relationship:

$$R_F = (K_1 + K_2 \cdot C_{ROOH}) \cdot \left(\frac{p_{O_2}^{Oil}}{K_3 + K_4 \cdot p_{O_2}^{Oil}} \right), \tag{3.40}$$

where the K_i are constants to be regarded as fitting parameters, and $p_{O_2}^{Oil}$ is the oxygen partial pressure, which, assuming that the solubilization process of oxygen into oil is governed by Hanry's law, is related to the oxygen concentration in the oil ($C_{O_2}^{Oil}$) through the relationship $p_{O_2}^{Oil} = \frac{C_{O_2}^{Oil}}{S_{O_2}^{Oil}}$, where $S_{O_2}^{Oil}$ is oxygen solubility in the virgin olive oil.

Hydroperoxides break down, giving rise to several secondary products (Labuza 1971). Many reactions are involved in this process; each of them is characterized by a particular mechanism and should be described by a specific equation. For the sake of simplicity, Del Nobile et al. (2003) assumed that at a given temperature the overall rate at which hydroperoxides decompose depends only on their concentration according to the following expression:

$$R_D = K_5 \cdot C_{ROOH}, \tag{3.41}$$

where K_5 is a constant to be regarded as a fitting parameter.

Substituting Eqs. 3.40 and 3.41 into Eq. 3.39 the following expression is obtained:

$$\frac{\partial C_{ROOH}}{\partial t} = (K_1 + K_2 \cdot C_{ROOH}) \cdot \left(\frac{p_{O_2}^{Oil}}{K_3 + K_4 \cdot p_{O_2}^{Oil}} \right) - K_5 \cdot C_{ROOH}. \qquad (3.42)$$

The mass balance on the oxygen dissolved in the bottled oil has the following expression:

$$\frac{\partial C_{O_2}^{Oil}}{\partial t} = \zeta \cdot \left[\frac{D_{O_2}^{Oil}}{r} \cdot \frac{\partial}{\partial r} \left(\frac{\partial C_{O_2}^{Oil}}{\partial r} \right) \right] - (K_1 + K_2 \cdot C_{ROOH}) \cdot \left(\frac{p_{O_2}^{Oil}}{K_3 + K_4 \cdot p_{O_2}^{Oil}} \right), \qquad (3.43)$$

where r is the spatial coordinate, $D_{O_2}^{Oil}$ is the oxygen diffusivity through the oil, and ζ is a constant equal to 1 for plastic containers and to 0 for glass containers. The term on the right-hand side of Eq. 3.43 enclosed in square brackets is related to the oxygen diffusive mass flux and was obtained by Del Nobile et al. (2003a) assuming that the oxygen diffusion and solubilization processes in the oil were governed by Fick's first law and Henry's law, respectively.

In the case of gas-permeable containers, like plastic containers, to evaluate the amount of oxygen permeating through the container wall, it is necessary to write the mass balance equation for the oxygen dissolved in the container wall, which, in the case investigated by Del Nobile et al. (2003a), has the following expression:

$$\frac{\partial C_{O_2}^{Polym.}}{\partial t} = \frac{1}{r} \cdot \frac{\partial}{\partial r} \left(r \cdot D_{O_2}^{Polym.} \cdot \frac{\partial C_{O_2}^{Polym.}}{\partial r} \right), \qquad (3.44)$$

where $C_{O_2}^{Polym.}$ is the concentration of the oxygen dissolved in the container wall, and $D_{O_2}^{Polym.}$ is the diffusivity of oxygen through the plastic container wall. As reported earlier, for gas diffusion through a glassy polymer $D_{O_2}^{Polym.}$ depends on the oxygen partial pressure according to Eq. 3.35 (Paul and Koros 1976), which can be rearranged as follow:

$$D_{O_2}^{Polym.} = D_{O_2}^{D} \cdot \left[\frac{1 + \dfrac{\left(\dfrac{D_{O_2}^{H}}{D_{O_2}^{D}} \right) \cdot \left(\dfrac{C_H' \cdot b}{k_D} \right)}{\left(1 + b \cdot p_{O_2}^{Polym.} \right)^2}}{1 + \dfrac{\dfrac{C_H' \cdot b}{k_D}}{\left(1 + b \cdot p_{O_2}^{Polym.} \right)^2}} \right], \qquad (3.45)$$

where $p_{O_2}^{Polym.}$ is the oxygen partial pressure in the container wall. At low oxygen partial pressure, as in the case investigated by Del Nobile et al. (2003a), Eq. 3.45 becomes

$$D_{O_2}^{Polym.} = D_{O_2}^{D} \cdot \left[\frac{1 + \left(\frac{D_{O_2}^{H}}{D_{O_2}^{D}} \right) \cdot \left(\frac{C_H' \cdot b}{k_D} \right)}{1 + \frac{C_H' \cdot b}{k_D}} \right]. \tag{3.46}$$

In this case $D_{O_2}^{Polym.}$ is constant, and hence Eq. 3.44 can be rewritten as follows:

$$\frac{\partial C_{O_2}^{Polym.}}{\partial t} = \frac{D_{O_2}^{Polym.}}{r} \cdot \frac{\partial}{\partial r} \left(r \cdot \frac{\partial C_{O_2}^{Polym.}}{\partial r} \right). \tag{3.47}$$

Interfacial conditions were imposed to ensure equilibrium conditions at the interface between oil and plastic. In fact, both the oxygen mass flow and the oxygen partial pressure of the juxtaposed substances were required to be equal at the interface:

$$\left[J_{O_2}^{Polym.} \right]_{int.} = \left[J_{O_2}^{Oil} \right]_{int.}, \tag{3.48}$$

$$\left[p_{O_2}^{Polym.} \right]_{int.} = \left[p_{O_2}^{Oil} \right]_{int.}, \tag{3.49}$$

where $\left[J_{O_2}^{Polym.} \right]_{int.}$ is the oxygen mass flux at the interface in the container wall, and $\left[J_{O_2}^{Oil} \right]_{int.}$ is the oxygen mass flux at the interface in the oil.

Equations 3.42, 3.44, and, only for gas-permeable containers, Eqs. 3.47, 3.48, and 3.49 form a set of differential equations that, using the proper initial and boundary conditions, were numerically solved by Del Nobile et al. (2003a) to predict the evolution of oxygen and hydroperoxides inside bottled oil during storage. The average hydroperoxide concentration (C_{ROOH}^{Av}) was obtained by averaging C_{ROOH} over the volume of the bottled olive oil.

The illustrated model was tested by the authors using three different bottles: (1) a commercially available glass bottle with a capacity of 500 cm^3, henceforth referred to as the sample VOO/Glass; (2) a commercially available PET bottle with an average wall thickness of 0.035 cm, a diameter equal to 6 cm, and a capacity of 480 cm^3, henceforth referred to as the sample VOO/PET; (3) a prototype bottle obtained by using an experimental material based on a starch/PCL blend with a wall thickness of 0.076 cm, a diameter of 5 cm, and a capacity of 200 cm^3, henceforth referred to as the sample VOO/Blend.

The following oxygen transport parameters were obtained by Del Nobile et al. (2003): $S_{O_2}^{oil} \left[\frac{cm^3(STP)}{cm^3 \cdot atm}\right] = 0.15$; $D_{O_2}^{oil} \left[\frac{cm^2}{s}\right] = 2.72 \times 10^{-6}$; $S_{O_2}^{PET} \left[\frac{cm^3(STP)}{cm^3 \cdot atm}\right] = 0.078$; $D_{eff}^{PET} \left[\frac{cm^2}{s}\right] = 8.8 \times 10^{-9}$; $S_{O_2}^{starch-PCL} \left[\frac{cm^3(STP)}{cm^3 \cdot atm}\right] = 0.081$; $D_{O_2}^{starch-PCL} \left[\frac{cm^2}{s}\right] = 1.26 \times 10^{-7}$.

To evaluate constants K_1, K_2, K_3, K_4, and K_5, Del Nobile et al. (2003a) fitted Eqs. 3.42 and 3.44 to data relative to the sample VOO/Glass.

The results of fitting parameters were: $K_1 = 8.52 \cdot 10^{-9} \frac{cm^3(STP)}{cm^3 \cdot s}$; $K_2 = 2.47 \cdot 10^{-10}$; $K_3 = 7.32 \cdot 10^{-3}$ atm; $K_4 = 1.02$; $K_5 = 3.28 \cdot 10^{-8} \frac{1}{s}$. As highlighted by Del Nobile et al. (2003a), although the model satisfactorily interpolated the experimental data, due to the limited number of available experimental data, the K_i values should only be regarded as a rough estimation of these parameters.

To assess the predictive ability of the mechanistic model, Del Nobile et al. (2003a) used the previously reported oxygen transport parameters and K_i parameters to predict the evolution of average hydroperoxide concentrations in the samples VOO/PET and VOO/Blend. Despite the restrictions imposed to derive the model, its predictive ability appeared to be quite satisfactory.

The major limitations to extending the proposed model to more general cases are the use of empirical Eqs. 3.40 and 3.41 to describe hydroperoxide formation and breakdown reactions and the fact that the oxygen mass flux in the bottle axial direction is considered negligible. Wherever these assumptions are satisfied, the previously described mechanistic model can be suitably adopted to interpolate or predict the quality decay kinetic of bottled oil.

Del Nobile et al. (2003a) also used the model to show the limit, in terms of performance, of the use of plastic containers and to evaluate the benefits arising from the use of innovative packaging solutions. In the examples reported below, a cylindrically shaped bottle with 8 cm diameter was used by the authors to run all simulations. The rate of hydroperoxide formation and breakdown was evaluated using the K_i values reported earlier, whereas the oxygen transport parameters were used, unless otherwise specified, to describe the diffusivity and solubilization of oxygen in the olive oil, PET, and starch/PCL blend. Figure 3.13a shows the predicted average concentration of hydroperoxides plotted against storage time. The curves refer to virgin olive oil bottled in plastic containers made of polymers that differ in the value of the oxygen diffusion coefficient. For comparative purposes, the curve representing the behavior of the sample VOO/Glass as predicted by the model is also reported in the same figure. Figure 3.13b also shows the predicted average concentration of hydroperoxides plotted against storage time, but for virgin olive oil bottled in PET containers differing in the thickness of the wall container. Also in this case the curve representing the predicted behavior of the sample VOO/Glass is reported for comparative purposes. From the data shown in the two graphs of Fig. 3.13 it emerges that by reducing the oxygen diffusivity of the plastic container it is possible to slow down the quality decay kinetic to that of the sample VOO/Glass. In contrast, the same result cannot be obtained by increasing the thickness of the plastic container. In fact, the amount of

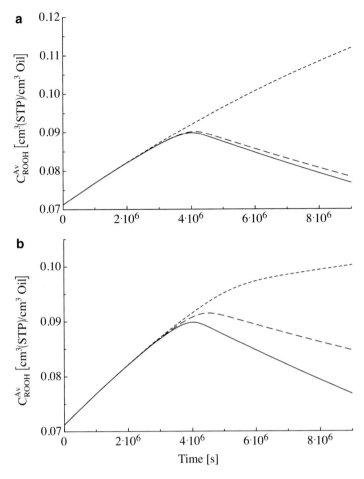

Fig. 3.13 Predicted average hydroperoxide concentration versus time. (**a**) (——) Glass; PET bottles with same thickness (0.35 cm) and different oxygen diffusion coefficient (− − −) D = $5 \cdot 10^{-8}$ cm²/s; (---) D = $8.8 \cdot 10^{-9}$ cm²/s. (**b**) (—) Glass; PET bottles with same oxygen diffusion coefficient ($8.8 \cdot 10^{-8}$ cm²/s) and different thickness (− − −) 0.55 cm; (---) 0.15 cm

oxygen that diffuses through the plastic container, causing an increase in the average hydroperoxide concentration, comes partially from outside the bottle and partially from the oxygen dissolved in the bottle wall. As the wall thickness increases, only the former term vanishes while the latter does not change substantially.

To demonstrate the benefits of using an oxygen scavenger, this mechanistic model was used by Del Nobile et al. (2003a) to predict the quality decay kinetic of virgin olive oil bottled in an innovative plastic container, the oxygen scavenger being uniformly dispersed in the container wall. In the presence of an oxygen scavenger, Eq. 1.2 becomes

$$\frac{\partial C_{O_2}^{\text{Polym.}}}{\partial t} = \frac{D_{O_2}^{\text{Polym.}}}{r} \cdot \frac{\partial}{\partial r} \left(r \cdot \frac{\partial C_{O_2}^{\text{Polym.}}}{\partial r} \right) - R_{\text{O.S.}}, \tag{3.50}$$

where $R_{\text{O.S.}}$ is the rate at which oxygen is consumed by the oxygen scavenger. In principle, $R_{\text{O.S.}}$ depends on the concentration of oxygen and oxygen scavenger in the plastic, temperature, and humidity. Since there are no data reported in the literature on the relationship between $R_{\text{O.S.}}$ and the physical quantities, the empirical equation used by Del Nobile et al. (2003) is reported as

$$R_{\text{O.S.}} = K_6 \cdot C_{O_2}^{\text{Polym.}} \cdot C_{\text{O.S.}}^{\text{Polym.}}, \tag{3.51}$$

where K_6 is a constant that depends on the temperature and humidity, and $C_{\text{O.S.}}^{\text{Polym.}}$ is the oxygen scavenger concentration in a plastic container. The authors set the value of K_6 equal to $1.58 \cdot 10^{-5} \ \frac{\text{cm}^3}{\text{cm}^3(\text{STP}) \cdot \text{s}}$ but the initial value of $C_{\text{O.S.}}^{\text{Polym.}}$ equal to $14 \ \frac{\text{cm}^3(\text{STP})}{\text{cm}^3}$.

If oxygen scavengers are used in the bottle, a lower average hydroperoxide concentration can be obtained. In fact, in the work cited, plastic bottles containing the oxygen scavenger differing in the oxygen diffusion coefficient were used. By comparing the trend of hydroperoxide evolution in the presence of oxygen scavengers to the behavior in glass, PET, or blend, a slower quality decay kinetic was found. Interestingly, by using oxygen scavengers in the starch/PCL blend, a useful container was obtained, with better performance than glass and characterized by a low environmental impact.

Figure 3.14a shows the predicted average hydroperoxide concentration plotted as a function of storage time for olive oil stored in glass bottles internally coated with a polymeric film in which oxygen scavengers are uniformly dispersed. The curves differ in the oxygen diffusivity of the polymeric coating. For comparative purposes, the curve representing the predicted behavior of the sample VOO/Glass is also given in the same figure. The coated bottles show a slower quality decay kinetic compared to glass packaging. Moreover, using a polymer with the highest oxygen diffusivity yields the slowest quality decay kinetic. In fact, for this type of bottle the oxygen scavenger is used to remove the oxygen dissolved in the bottled oil. Therefore, by increasing the oxygen diffusivity of the polymeric coating, the rate at which the oxygen dissolved in the oil diffuses into the polymeric coating increases, leading to a slowdown in the quality loss kinetic. Figure 3.14b shows the predicted average hydroperoxide concentration plotted as a function of storage time for virgin olive oil bottled in PET containers. The reported curves refer to virgin olive oil with a different initial concentration of dissolved oxygen. For comparative purposes, the behavior of the samples VOO/Glass and VOO/PET as predicted by the mechanistic model illustrated earlier is also reported. As would be expected, reduction of the amount of oxygen dissolved in the oil led to a slowdown in the quality decay. In particular, when the oxygen concentration was equal to 10 % of

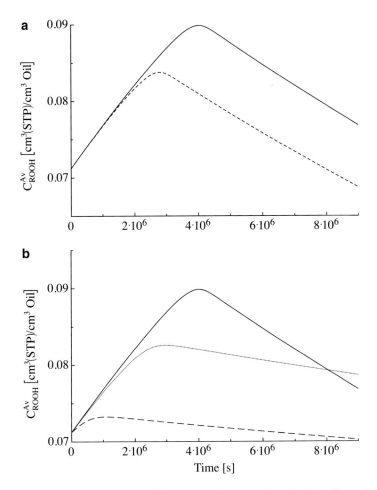

Fig. 3.14 Predicted average hydroperoxide concentration versus time for olive oil stored in glass bottles internally coated with a polymeric film in which oxygen scavengers are uniformly dispersed. (**a**) (——) Glass; PET with oxygen diffusion coefficient (— — —) $D = 8.8 \cdot 10^{-9}$ cm²/s. (**b**) (——) Glass sample; PET bottles with same oxygen diffusion coefficient ($8.8 \cdot 10^{-9}$ cm²/s), same thickness (0.35 cm), and different initial concentration of dissolved oxygen (— — —) 90 %; (·····) 50 %

the equilibrium concentration (i.e., oxygen concentration in virgin olive oil equilibrated at an oxygen partial pressure of 0.02 atm), the slowest decay kinetic among those examined by Del Nobile et al. (2003a) was obtained.

References

Barrer RM, Barrie JA, Slater J (1958) Sorption and diffusion of ethyl cellulose. III: comparison between ethyl cellulose and rubber. J Polym Sci 27:177–197

Bell LN, Labuza TP (2000) Moisture sorption: practical aspects of sorption isotherm measurement and use, 2nd edn. American Association of Cereal Chemists, St. Paul, p 57

Blekas G, Tsimidou M, Boskou D (1995) Contribution of a-tocopherol to olive oil stability. Food Chem 52:289–294

Chern RT, Koros WJ, Sanders ES, Yui R (1983) Second component. Effects in sorption and permeation of gases in glassy polymers. Membr Sci 15:157–169

Conte A, Scrocco C, Brescia I, Del Nobile MA (2009) Packaging strategies to prolong the shelf life of minimally processed lampascioni (*Muscari comosum*). J Food Eng 90:199–206

Del Nobile MA (2001) Packaging design for potato chips. J Food Eng 47:211–215

Del Nobile MA, Mensitieri G, Nicolais L, Masi P (1997) The influence of the thermal history on the shelf life of carbonated beverages bottled in plastic containers. J Food Eng 34:1–13

Del Nobile MA, Ambrosino ML, Sacchi R, Masi P (2003a) Design of plastic bottles for packaging of virgin olive oil. J Food Sci 68:170–175

Del Nobile MA, Buonocore GG, Limbo S, Fava P (2003b) Shelf life prediction of cereal based dry foods packed in moisture sensitive films. J Food Sci 68:1292–1300

Del Nobile MA, Conte A, Cannarsi M, Sinigaglia M (2008a) Use of biodegradable films for prolonging the shelf life of minimally processed lettuce. J Food Eng 85:317–325

Del Nobile MA, Sinigaglia M, Conte A, Speranza B, Scrocco C, Brescia I, Bevilacqua A, Laverse J, La Notte E, Antonacci D (2008b) Influence of postharvest treatments and film permeability on quality decay kinetics of minimally processed grapes. Postharvest Biol Technol 47:389–396

Del Nobile MA, Conte A, Scrocco C, Brescia I, Speranza B, Sinigaglia M, Perniola R, Antonacci D (2009) A study on quality loss of minimally processed grapes as affected by film packaging. Postharvest Biol Technol 51:21–26

Fava P, Limbo S, Piergiovanni L (2000) Shelf-life modeling of moisture sensitive food products. Industrie Alimentari 32:121–128

Fenelon PJ (1973) Prediction of pressure loss in pressurized plastic containers. Polym Eng Sci 13:440–451

Frankel EN (1998) Lipid oxidation. The Oily Press, Dundee, p 303

Heiss R (1958) Shelf life determination. Modern Packaging 31:119–127

Iglesias HA, Viollaz P, Chirife J (1979) Technical note: a technique for predicting moisture transfer in mixtures of packaged dehydrated foods. J Food Sci 14:89–93

Koros WJ (1980) Model for sorption of mixed gases in glassy polymers. J Polym Sci 18:981–992

Koros WJ, Hellums MW (1990) Encyclopedia of polymer science and engineering. Wiley, New York, pp 724–802, Supplement Volume

Koros WJ, Paul DR (1978) CO_2 sorption in poly(ethylene terephthalate) above and below the glass transition. J Polym Sci 16:1947–1963

Labuza TP (1971) Kinetics of lipid oxidation in foods. CRC Crit Rev Food Technol 2:355–405

Labuza TP, Contreras-Medellin R (1981) Prediction of moisture protection requirements for foods. Cereal Foods World 26:335–343

Lucera A, Costa C, Mastromatteo M, Conte A, Del Nobile MA (2010) Influence of different packaging systems on fresh-cut zucchini (*Cucurbita pepo*). Innov Food Sci Emerg Tec 11:361–368

Masi P, Paul DR (1982) Modelling gas transport in packaging applications. J Membr Sci 12:137–151

Michaels AS, Vieth WR, Barrie JA (1963) Solution of gases in polyethylene ter-ephtalate. J Polym Sci 34:1–12

Myers AW, Meyer JA, Rogers CE, Stannett V, Szwarc M (1961) Studies in the gas and vapor permeability of plastic films and coated papers. IV: The permeability of water vapor. Tappi 44:58–64

Papadopulos G, Boskou D (1991) Antioxidant effect of natural phenols in olive oil. J Am Oil Chem Soc 68:669–678

Paul DR (1979) Gas sorption and transport in glassy polymers. Bel: Bunsenges. Phys Chem 83:294–302

Paul DR, Koros WJ (1976) Effect of partially immobilizing sorption on permeability and diffusion time lag. J Polym Sci 14:675–685

Press WH, Flannery BP, Teukolsky SA, Vetterling WT (1989a) Numerical recipes in Pascal. Cambridge University Press, Cambridge, pp 602–607

Press WH, Flannery BP, Teukolsky SA, Vetterling WT (1989b) Numerical recipes in pascal. Cambridge University Press, Cambridge, p 122

Quast DG, Karel M (1972) Computer simulation of storage life of foods undergoing spoilage by two interacting mechanisms. J Food Sci 37:679–683

Quast DG, Karel M, Rand M (1972) Development of a mathematical model for oxidation of potato chips as a function of oxygen pressure, extent of oxidation and equilibrium relative humidity. J Food Sci 37:673–678

Rai DR, Tyagi SK, Jha SN, Mohan S (2008) Qualitative changes in the broccoli (*Brassica oleracea var. italica*) under modified atmosphere packaging in perforated polymeric film. J Food Sci Technol 45:247–250

Rogers C, Meyer J A, Stannett V, Szwarc M (1956) Studies in the gas and vapor permeability of plastic films and coated papers. I: Determination of the permeability constant. Tappi 39: 737–741

Saravacos GD (1986) Mass transfer properties of foods. In: Rao MA, Rizvi SSH (eds) Engineering properties of foods. Marcel Dekker, New York, p 161

Satue MT, Huang SW, Frankel EN (1995) Effect of natural antioxidants in virgin olive oil on oxidative stability of refined, bleached, and deodorized olive oil. J Am Oil Chem Soc 72:1131–1137

Tsimidou M, Papadopoulos G, Boskou D (1992) Phenolic compounds and stability of virgin olive oil. Part I. Food Chem 45:141–144

Tubert AH, Iglesias HA (1985) Water sorption isotherms and prediction of moisture gain during storage of packaged cereal crackers. Lebensm Wiss Technol 19:365–368

Vermeiren L, Devlieghere F, van Beest M, de Kruijf N, Debevere J (1999) Developments in the active packaging. Trends Food Sci Technol 10:77–86

Vieth WR, Tam PM, Michaels AS (1966) Dual sorption mechanism in glassy poly-styrene. J Colloid Interface Sci 22:360–370

Wolf W, Spiess WEL, Jung G (1985) Standardization of isotherm measurements (COST-PROJECT 90 and 90bis). In: Simatos D, Multon JL (eds) Properties of water in foods, vol 90, NATO ASI series, series E: applied sciences. M. Nijhoff, Dordrecht/Boston, p 661

Yasuda H, Stannett V (1975) Permeability coefficients. In: Brandrup J, Immergut EH (eds) Polymer handbook. Wiley, New York

Part II
Low-Environmental-Impact Active Packaging

This section will consider the most relevant research dealing with green polymers with active properties. The modification of a polymeric matrix intended for food applications aimed at preventing microbial contamination or spoilage growth is highly desirable. Thus, the interest in developing materials with antimicrobial properties for biomedical, food, and cosmetic applications is very high. The active agents should combine the potential bactericidal or fungicidal effect to characteristics of environmental safety, low toxicity, and low cost. Impregnated packaging films and coatings that release natural compounds to extend food shelf life and active packaging systems where the preservative is completely immobilized to polymeric anchor groups and acts directly from the material without being released into the packaged foodstuff will be discussed. Due to the growing interest in environmental pollution, attention will be also focused on active films produced by renewable sources. In this context, numerous research efforts have been made to develop active systems intended for food-packaging applications based on renewable polymeric materials. The incorporation of stiffening material (biofibers) in protein-based film-forming solution will also be presented and analyzed using a response surface methodology analysis.

Chapter 4
Different Approaches to Manufacturing Active Films

4.1 Introduction

Active packaging is a novel typology of materials for food applications that has great industrial and research relevance. These films have additional functions compared to traditional passive systems, which generally have the sole utility of passively protecting packaged products from external detrimental phenomena. With active materials it is possible to increase substantially the shelf life (SL) of packaged products by acting on specific mechanisms that bring about their unacceptability. Antimicrobial or antioxidant packaging, or systems that scavenge oxygen or ethylene from the package headspace of course represent the most diffuse and interesting active packaging solutions (Coma 2008). Even though dipping or spraying of these compounds on food can improve its quality, direct application on the product is generally not recommended due to the potential diffusion of active substances in the food matrix (Appendini and Hotchkiss 1997, 2001). The use of active packaging for food storage could represent a more suitable system to allow for a slow migration of active agents from polymeric material to food surface, thereby making it possible to maintain the quality of the packaged food (Joerger 2007). Promising active packaging systems are based on the incorporation of antimicrobial substances in the packaging material to control undesirable growth of microorganisms on food surfaces (Lopez-Rubio et al. 2006). Antimicrobial compounds embedded in polymers can act based on two different mechanisms. The preservative can be covalently immobilized into the polymer matrix and act directly from the film when the food is brought into contact with the active material. Under the second mechanism, the preservative is embedded into the matrix in the dry state. When the active material is brought into contact with a moist food or a liquidlike food, the preservative is released from the material and acts directly on the food. The choice of active system depends on various factors; the nature of the active compound, the type of polymeric material, and the food to be protected represent the main elements to be taken into account when creating a suitable device (Fernandez et al. 2008). For example, the inclusion of an

M.A. Del Nobile and A. Conte, *Packaging for Food Preservation*,
Food Engineering Series, DOI 10.1007/978-1-4614-7684-9_4,
© Springer Science+Business Media New York 2013

agent in a film could result in a lower effectiveness than that recorded after its application on the film surface. However, agents bound to the film surface are likely to be limited to enzymes or other proteins because their molecular structure must be large enough to retain activity on the microorganism cell wall while being bound to the plastic (Quintavalla and Vicini 2002).

The components of active antimicrobial systems can be either organic or inorganic. Inorganic systems are based on metal ions such as silver, silver zeolite particles, copper, and platinum, which are approved as additives in food contact polymers, especially in the USA and Japan (Brody 2001; Cutter 2002). Organic acids, bacteriocins, enzymes, and spice extracts have been studied as natural, organic food preservatives and alternatives to synthetic compounds because of their ability to prolong the SL of packaged food and for their potential safety for humans (Lee et al. 2003; Vermeiren et al. 2002). In the case of immobilized systems, a non-food-grade antimicrobial substance could also be utilized due to the absence of a diffusing mass transfer. The number of preservatives with potential antimicrobial properties that could be used in these types of packaging is very high (Suppakul et al. 2003). However, the limitation of such a system is the direct contact between the packaging and the food (Coma 2008). Ionic bonding of antimicrobials onto polymers allows slow release into the food. However, diffusion to the product is less of a concern when the antimicrobial is covalently bonded to the polymer, unless conditions within the product promote reactions such as hydrolysis. This may occur, for example, during the heating of a food with a high acid content (Appendini and Hotchkiss 2002).

Traditional packaging materials derived from petroleum oil are neither readily recyclable nor environmentally sustainable and impose a number of health risks associated with, for instance, the migration of harmful additives. Thus, it would be highly desirable to develop biodegradable (including edible) and sustainable matrices capable of safe and long SL integration of bioactive substances (Siracusa et al. 2008). This chapter surveys antimicrobial packaging materials with immobilized active compounds. The potential of these technologies is evaluated with a view to extending the SL and assure the innocuousness and preservation of fresh food.

4.2 Immobilization by Chemical Retention

Both ionic and covalent immobilization requires the presence of functional groups on both the antimicrobial molecule and the polymer. Examples of antimicrobials with functional groups are peptides, enzymes, polyamines, and organic acids. On the other hand, polymers with functional groups more commonly used for food packaging are ethylene vinyl acetate (EVA), ethylene methyl acrylate (EMA), ethylene metacrylic acid (EMAA), ionomer, nylon, polyvinylidene chloride (PVdC), ethylene vinyl alcohol (EVOH)/polyethylene (PE) copolymer, and polystyrene (PS) (Appendini and Hotchikiss 2002). Among the biomaterials, carrageenan (Van de Velde et al. 2002), chitosan (Agulló et al. 2003; Chang and Juang 2005), gelatin (Nagatomo et al. 2005), alginate (Roy and Gupta 2004), polylactic acid, and polyglycolic acid (Lazzeri et al. 2005) are very promising materials for the immobilization of enzymes due to their natural, biodegradable, biocompatibile, nontoxic, and ready availability.

The most important problem that one encounters with respect to covalent immobilization is the choice of a suitable spacer molecule [dextrans, polyethyleneglycol (PEG), ethylenediamine, and polyethylencimine] that is able to bind the bioactive agent to the polymer surface, thereby avoiding a reduction in antimicrobial activity that might result from immobilization (Conte et al. 2006a). An effort to increase the activity of an immobilized agent was carried out with naringinase, an enzyme that reduces juice bitterness by hydrolysis of naringine (Del Nobile et al. 2003; Soares and Hotchkiss 1998). The protection of active sites during film formation and the incorporation of dendrites to increase the surface area of the supports were tested with success.

The binding of an agent to the surface of a package would require a molecular structure large enough to retain activity on the microbial cell wall even though the agent is bound to the plastic. According to Quintavalla and Vicini (2002), such agents are likely to be limited to enzymes or other antimicrobial proteins. The covalent union between the polymer and the enzyme is based on the chemical activation of a surface and the attachment of nucleophilic groups of the proteins. This union transforms the polymer into a stable carrier that, in the best case, does not leach the protein to the surroundings, thereby avoiding the arrival of undesirable compounds to a potential food matrix that has been put in contact with the activated polymer (Goddard and Hotchkiss 2007). Lysozyme, chitinase, glucose oxidase, βeta-galactosidase, lactoferrin, lactase sulfhydril oxidase, and bile-salt-stimulated lipase are the most diffuse enzymes covalently immobilized onto different support materials (Appendini and Hotchkiss 2002; Goddard et al. 2007; Fernandez et al. 2008; Vartiainen et al. 2005). Due to the presence of amino and carboxylic groups, peptides were also used as valid candidates for covalent immobilization (Appendini and Hotchkiss 2001). The advantage is that the peptide was built directly on the resin by protecting the amino acid functional groups. Via covalent attachment to the sole histidine amino acid, the immobilization of hen egg white lysozyme onto inert soluble PS beads was also achieved. This single-point immobilization may minimize the stearic interference between the support and the immobilized enzyme (Wu and Daeschel 2007). Chemical immobilization of bacteriocins was also investigated. Scannell et al. (2000) studied the immobilization of various bacteriocins on PE/polyamide materials, showing that the immobilized nisin maintained its activity for 3 months in various storage temperatures, even if a certain decrease occurred in the activity. To protect nisin from denaturation, Millette et al. (2007) elaborated nisin covalently linked to activate alginate beads to be used in meat products. The alginate activation was carried out with sodium periodate to generate dialdehyde residues that, following reaction with nisin amine groups, formed a Schiff base linkage. After 14 days of storage, in ground beef samples containing active beads, a substantial reduction in microbial count was observed, demonstrating the effectiveness of the bioactive packaging.

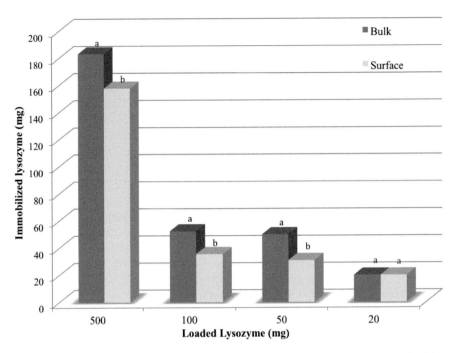

Fig. 4.1 Amounts of immobilized lysozyme (mg) determined after washing films obtained by bulk and surface immobilization. Statistically significant differences ($p < 0.05$) between samples are identified by different letters

The cross-linking technique to chemically bind active compounds to a polymer involves the use of chemical compounds often unsuited for food applications (Conte et al. 2006a, 2007; Del Nobile et al. 2003). A comparison between bulk and surface immobilization on polyvinyl-alcohol (PVOH) films, obtained by casting technique, was carried out by Conte et al. (2006a). Lysozyme was immobilized to PVOH by spraying glutaraldehyde onto the polymer surface or by mixing the enzyme and the chemical compound in the film forming solution. After washing of the films and monitoring of the unbound lysozyme released, the immobilized enzyme was quantified (Fig. 4.1). The amount of lysozyme bound to the polymeric matrix increased as the loaded lysozyme/binding agent ratio decreased. The authors ascribed the difference between the two types of film to the fact that glutaraldehyde can act both as binding agent for lysozyme and as cross-linking agent for a polymeric matrix. For this reason, in the case of bulk immobilization, glutaraldehyde was involved to a great extent in these two reactions compared to surface immobilization. As a consequence, the antimicrobial effectiveness of both active systems increased, with a linear dependence, as the amount of immobilized lysozyme increased; however, the activity of the films with surface immobilization was higher than bulk films, with lysozyme being active by contact. In the case of surface immobilized lysozyme, the amount of enzyme available to microbial cells was much higher than in the other films, suggesting that surface immobilization is more desirable for packaging purposes compared to a bulk system containing the same amount of active agent. Due to the obtained results, in a subsequent work, the same authors tested with success lysozyme immobilized onto a PVOH surface against *Alicyclobacillus acidoterrestris* in apple juice (Conte et al. 2006b). As was the case earlier, the active films were obtained by spraying a water solution of lysozyme (2 %), with a proper amount of glutaraldehyde, onto the PVOH surface in order to uniformly distribute the antimicrobial compound and the bonding agent; glacial acetic acid was also added as a reaction catalyst. The microbiological tests were performed on various volumes of juice, ranging from 150 to 600 ml, previously inoculated with both a single strain and a culture cocktail of microbial suspension. The authors found that the developed active film maintained its efficacy at different film surface/juice volume ratios. The same microbial tests were also conducted on viable spores of *A. acidoterrestris*, thereby demonstrating an enhanced efficacy of the active PVOH in inhibiting viable spores.

To overcome the problem of toxic compounds to activate the polymer surface, mineral supports, such as inorganic nanocarriers, could be added to biopolymer materials to promote enzyme binding by cation exchange, physical adsorption, or ionic binding. Alpha-amylase, glucoamylase, and invertase were immobilized through adsorption and grafting onto clays generally used as nanofillers in polymers to reinforce mechanical and barrier properties to low molecular weight compounds of packaging films (Gopinath and Sugunan 2007).

4.3 Immobilization by Polymer Surface Modification

To prevent the transfer or migration of antimicrobial substances from polymer to food, modification of polymers has been also investigated as a means to render surfaces antimicrobial (Haynie et al. 1995). Modifying surface composition of polymers by electron irradiation in such a way that the surface contains amine groups has also been shown to exhibit antimicrobial activity that inactivates microorganisms by contact. Exposure of nylon yarn or fabric to light from an excimer laser in air caused an apparent conversion of amides to amines, which are still bound to the polymer chain and were proved active against various spoilage and pathogen microorganisms (Shearer et al. 2000). UV irradiation, like that produced by a UV excimer laser at an appropriate wavelength, was also used to oxidize O_2, previously absorbed on a modified surface layer, to O_3 by photochemical means (Rooney 1995). The authors proved that even a small amount of the formed O_3 desorbed from the polymer matrix to the interior of the package was sufficient to inhibit microbial growth without direct contact between food and antimicrobial film.

Not only electron irradiation or UV treatments can be used to modify film surfaces: plasma treatments are under development as well (Favia et al. 2005). Low-temperature plasma modification techniques, in deposition and grafting modes, are widely used to modify the surface chemistry and properties of materials for microelectronics, packaging, textiles, biomaterials, and many other industrial applications (d'Agostino et al. 2005). Plasma modification processes make it possible to produce surfaces characterized by different types of functionalities (e.g., COOH, OH, NH_2), which can be used as anchor groups for the immobilization of various molecules at surfaces. Plasma-treated surfaces have been largely and successfully employed for the immobilization of various compounds to be used for biomedical and sensor applications (Dai et al. 2000). The possibility of immobilizing a wide range of molecules with conventional chemical organic reactions also opens the door to many possible applications for developing active packaging systems. Moreover, the technique gained considerable attention in the food packaging area as an effective method to improve the adhesion, sealability, and wettability of polymers (Gancarz et al. 2003). Low-pressure plasma processes (10^{-2}–10 Torr) allow surface chemical and physical modifications of the topmost layers of materials with no alteration of the bulk. Conte et al. (2008) tested the antimicrobial activity of PE films, modified by means of plasma processes, followed by the chemical surface immobilization of lysozyme. For the purposes of the work, PE films were grafted with O-containing chemical groups by means of radio frequency (RF) glow discharges fed with H_2O vapors, in order to activate the polymer surface for the immobilization of lysozyme from water solutions. To reduce the mobility of the groups grafted later on, and to limit considerably the aging of the resulting surface, the films were pretreated in RF glow discharges fed with H_2 to crosslink their surfaces before grafting O-groups with H_2O discharges. Then, plasma-treated PE films were reacted in lysozyme solutions at different concentrations in order to achieve immobilization at the film surface. The samples obtained in this way were thoroughly rinsed in distilled water to remove any adsorbed/unbound lysozyme and analyzed by means of X-ray Photoelectron Spectroscopy (XPS) to check whether the enzyme was immobilized on plasma-treated substrates. A drastic change in the surface chemical composition was observed compared to the untreated samples. In particular, after immersion in the lysozyme solution, the N/C and O/C surface ratios of the samples were found to have considerably increased, and approximately 1 % of sulfur was detected. The amino acidic sequences in the enzyme structure confirmed the immobilization. The most accredited hypothesis to explain the lysozyme immobilization was based on the formation of ionic and electrostatic bonds, rather than covalent links, between the polar functional groups (e.g., $-NH_2$) of the enzyme and O-groups grafted at the surface of the PE films after the plasma processes. Surface analysis data also revealed no correlation between the surface composition of plasma-treated samples and the density of the immobilized lysozyme. At the same time, there was no clear effect of the increased power of plasma treatment on the antimicrobial activity of the processed films, even if all treated PE films resulted in effective antimicrobial surfaces. Bearing in mind that further work is still necessary to fully understand both the nature of the surface-lysozyme binding interactions and the mechanism of antibacterial action of the immobilized enzyme, the authors concluded that the plasma processes proved to be effective at producing enzyme-loaded antimicrobial surfaces that might find practical application in the field of active food packaging.

References

Agullò E, Rodriguez MS, Ramos V, Albertengo L (2003) Present and future role of chitin and chitosan in food. Macromol Biosci 3:521–530

Appendini P, Hotchkiss JH (1997) Immobilization of lysozyme on food contact polymers as potential antimicrobial films. Packag Technol Sci 10:271–279

Appendini P, Hotchkiss JH (2001) Surface modification of poly(styrene) by the attachment of an antimicrobial peptide. J Appl Polym Sci 81:609–616

Appendini P, Hotchkiss JH (2002) Review of antimicrobial food packaging. Innov Food Sci Emerg Technol 3:113–126

Brody AL (2001) What's active in active packaging. Food Technol 55:104–106

Chang MY, Juang RS (2005) Activities, stabilities, and reaction kinetics of three free and chitosaneclay composite immobilized enzymes. Enzyme Microb Technol 36:75–82

Coma V (2008) Bioactive packaging technologies for extended shelf life of meat-based products. Meat Sci 78:90–103

Conte A, Buonocore GG, Bevilacqua A, Sinigaglia M, Del Nobile MA (2006a) Immobilization of lysozyme on polyvinylalcohol films for active packaging applications. J Food Prot 4:866–870

Conte A, Sinigaglia M, Del Nobile MA (2006b) Antimicrobial effectiveness of lysozyme immobilized on poly-vinyl-alcohol-based film against *Alicyclobacillus acidoterrestris*. J Food Prot 4:861–865

Conte A, Buonocore GG, Sinigaglia M, Del Nobile MA (2007) Development of immobilized lysozyme based active film. J Food Eng 78:741–745

Conte A, Buonocore GG, Sinigaglia M, Lopez LC, Favia P, D'Agostino R, Del Nobile MA (2008) Antimicrobial activity of immobilized lysozyme on plasma-treated polyethylene films. J Food Prot 71:119–125

Cutter CN (2002) Microbial control by packaging: a review. Crit Rev Food Sci Nutr 42:151–161

d'Agostino RP, Favia P, Oehr C, Wertheimer M (2005) Low temperature plasma processing of materials: past, present and future. Plasma Processes Polym 2:7–15

Dai L, St John HAW, Bi J, Zientek P, Chatelier RC, Griesser HJ (2000) Biomedical coatings by the covalent immobilization of polysaccharides onto gas-plasma-activated polymer surfaces. Surf Interface Anal 29:46–55

Del Nobile MA, Piergiovanni L, Buonocore GG, Fava P, Puglisi ML, Nicolais L (2003) Naringinase immobilization in polymeric films intended for food packaging applications. J Food Sci 68:2046–2049

Favia P, Milella A, Iacobelli L, D'Agostino R (2005) Plasma pre-treatments and treatments on poly-tetrafluoroethylene for reducing the hydrophobic recovery. In: D'Agostino R, Favia P, Wertheimer RM, Oehr C (eds) Plasma processes and polymers. Wiley-VCH, Weinheim, pp 271–280

Fernandez A, Cava D, Ocio MJ, Lagaron JM (2008) Perspective for biocatalysts in food packaging. Trends Food Sci Technol 19:198–206

Gancarz I, Bryjak J, Pozniak G, Tylus W (2003) Plasma modified polymers as a support for enzyme immobilization II Amines plasma. Eur Polym J 39:2217–2224

Goddard JM, Hotchkiss JH (2007) Polymer surface modification for the attachment of bioactive compounds. Prog Polym Sci 32:698–725

Goddard JM, Talbert JN, Hotchkiss JH (2007) Covalent attachment of lactase to low-density polyethylene films. J Food Sci 72:6–41

Gopinath S, Sugunan S (2007) Enzymes immobilized on montmorillonite K10: effect of adsorption and grafting on the surface properties and the enzyme activity. Appl Clay Sci 35:67–75

Haynie SL, Crum GA, Doele BA (1995) Antimicrobial activities of amphiphilic peptides covalently bonded to a water insoluble resin. Antimicrob Agents Chemother 39:301–307

Joerger RD (2007) Antimicrobial films for food applications: a quantitative analysis of their effectiveness. Packag Technol Sci 20:231–273

Lazzeri L, Cascone MG, Quiriconi S, Morabito L, Giusti P (2005) Biodegradable hollow microfibres to produce bioactive scaffolds. Polym Int 54:101–107

Lee CH, An DS, Park HJ, Lee DS (2003) Wide spectrum antimicrobial packaging materials incorporating nisin and chitosan in the coating. Packag Technol Sci 16:99–106

Lopez-Rubio A, Almenar E, Hernandez-Munoz P, Lagaron JM, Catala R, Gavara R (2006) Overview of active polymer-based packaging technologies for food applications. Food Rev Int 20:357–387

Millette M, Le Tien C, Smoragiewicz W, Lacroix M (2007) Inhibition of Staphylococcus aureus on beef by nisin-containing modified alginate films and beads. Food Control 18:878–884

Nagatomo H, Matsushita Y, Sugamoto K, Matsui T (2005) Preparation and properties of gelatin-immobilized b-glucosidase from Pirococcus furiosus. Biosci Biotechnol Biochem 69:128–136

Quintavalla S, Vicini L (2002) Antimicrobial food packaging in meat industry. Meat Sci 62:373–380

Rooney ML (1995) Active packaging in polymer films. In: Rooney ML (ed) Active food packaging. Blackie Academic & Professional, Glasgow, pp 74–110

Roy I, Gupta MN (2004) Hydrolysis of starch by a mixture of glucoamylase and pullulanase entrapped individually in calcium alginate beads. Enzyme Microb Technol 34:26–32

Scannell AGM, Hill C, Ross RP, Marx S, Hartmeier W, Arendt EK (2000) Development of bioactive food packaging materials using immobilised bacteriocins Lacticin 3147 and Nisaplin. Int J Food Microbiol 60:241–249

Shearer AEH, Paik JS, Hoover DG, Haynie SL, Kelley MJ (2000) Potential of an antibacterial ultraviolet-irradiated nylon film. Biotechnol Bioeng 67:141–146

Siracusa V, Rocculi P, Romani S, Dalla Rosa M (2008) Biodegradable polymers for food packaging: a review. Trends Food Sci Technol 19:634–643

Soares NFF, Hotchkiss JH (1998) Naringinase immobilization in packaging films for reducing naringin concentration in grapefruit juice. J Food Sci 63:61–65

Suppakul P, Miltz J, Sonneveld K, Bigger SW (2003) Active packaging technologies with an emphasis on antimicrobial packaging and its applications. J Food Sci 68:408–420

Van de Velde F, Lourenco ND, Pinheiro HM, Bakker M (2002) Carrageenan: a food-grade and biocompatible support for immobilization techniques. Adv Synth Catal 344:815–835

Vartiainen J, Ratto M, Paulussen S (2005) Antimicrobial activity of glucose oxidase-immobilized plasma-activated polypropylene films. Packag Technol Sci 18:243–251

Vermeiren L, Devlieghere F, Debevere J (2002) Effectiveness of same recent antimicrobial packaging concepts. Food Addit Contam 19:163–171

Wu Y, Daeschel MA (2007) Lytic antimicrobial activity of hen egg white lysozyme immobilized to polystyrene beads. Food Microbiol Saf 72:369–374

Chapter 5
Bio-Based Packaging Materials for Controlled Release of Active Compounds

5.1 Introduction

Release packaging systems offer significant potential for extending the shelf life (SL) of foods by releasing antioxidants and antimicrobials over time, to replenish consumed active components originally present. Generally, active compounds are incorporated into food formulation. A limitation of this traditional method is that once the active compounds are consumed in reaction, the protection ceases and the quality of the food degrades at an increasing rate. Another limitation of the release systems is the inability to selectively target the food surface where most spoilage reactions occur; as a result, an extra amount of active compound is also unnecessarily added to the food product. Controlled release packaging can overcome these two limitations by continuously replenishing active compounds to the food surface, compensating for the consumption or degradation of active molecules, so that a predetermined concentration of compound is maintained in the food to achieve a desired SL (Mastromatteo et al. 2010). The incorporation of an antimicrobial substance into a food packaging system can take several approaches. One is to put the antimicrobial material into the film by adding it to the extruder when the film or the coextruded film is produced. The disadvantage of producing films containing active agents is the poor cost effectiveness since antimicrobial material not exposed to the surface of a film is generally not totally available to the antimicrobial activity. An alternative is to apply the antimicrobial additive in a controlled matter; for example, it can be incorporated into the food-contact layer of a multilayer packaging system (Quintavalla and Vicini 2002). Several factors, such as the chemical nature of film/coating, the process conditions, or the residual antimicrobial activity, must be taken into account in the design or modeling of the antimicrobial film (Suppakul et al. 2003b).

The migration of a substance may be achieved by direct contact between food and packaging material or through gas-phase diffusion from packaging layer to food surface (Coma 2008; Quintavalla and Vicini 2002). The theoretical advantage of volatile antimicrobials is that they can penetrate the bulk matrix of the food and

M.A. Del Nobile and A. Conte, *Packaging for Food Preservation*,
Food Engineering Series, DOI 10.1007/978-1-4614-7684-9_5,
© Springer Science+Business Media New York 2013

the polymer does not necessarily need to be in direct contact with the packaged food. Antimicrobial vapors or gases are appropriate for applications where contact between the required portions of the food and the packaging does not occur, as in ground beef or cut produce. Among the release device categories, a further division can be made between controlled and uncontrolled release systems. Even though uncontrolled delivery packages intended for food applications are the most abundant, controlled release systems are of industrial relevance due to their ability to prevent sensory or toxicological problems or system inefficiency caused by a too high or a too low concentration of a delivered substance. It is a challenge to apply the knowledge gained from pharmaceutical applications of drugs to the food industry. In fact, the design of controlled release systems is usually intended to optimize therapeutic regimens by providing for the slow and continuous delivery of a drug over the entire dosing interval while also providing greater patient compliance and convenience (Cooney 1972). Mathematical modeling of mass transfer is necessary for an in-depth understanding and optimization of active systems. Although data about numerical methods to describe controlled release mechanisms exist, most studies have been carried out using food-simulating systems, without taking into account the potential effects of such a release system on real food cases. Even though appropriate polymer films and information about factors controlling active compound release are available, release rates attainable with many current packagings are indeed too fast or too slow for practical use. Moreover, the development of a controlled migration of active compounds has been hampered by a lack of information about the real benefits deriving from the application to food of controlled release packaging systems compared to direct addition of active compounds or uncontrolled release systems. Reducing the amount of active compound in food may also provide improved flavor quality since many additives give a burning or off-flavor (Mastromatteo et al. 2010).

The release rate of an active compound from packaging to food can be realized by means of two mechanisms: a reservoir system and a swelling system. The former consists of an active agent contained within a rate-controlling microporous, macroporous, or nonporous barrier. The release rate from a reservoir system depends on the thickness, the area, and the permeability of the barrier. In a reservoir containing an excess of active agent, the constant release rate follows a zero-order kinetic. According to Pothakamury and Barbosa (1995), the principal steps in the release of an active ingredient from a reservoir system are (1) the diffusion of the active agent within the reservoir, (2) the dissolution or partitioning of the active agent between the reservoir carrier fluid and the barrier, (3) the diffusion through the barrier and partitioning between the barrier and the elution medium (i.e., the surrounding food), and (4) transport away from the barrier surface into the food. The rate-limiting step in the release of the active ingredient is diffusion through the polymeric barrier. On the other hand, in a swelling-controlled system, the active agent, dissolved or dispersed in a polymeric matrix, is unable to diffuse to any significant extent within the matrix because of its low diffusion coefficient. When the polymer matrix is placed in a thermodynamically compatible medium, the polymer swells owing to absorption of fluid (penetrant) from the medium. The active agent diffusion coefficient in the swollen part of the matrix increases and then diffuses out. In a diffusion-controlled system the matrix is assumed to be stationary during the release process, whereas in swelling-controlled systems, the membrane undergoes a transition from a glassy to a rubbery state upon interaction with the penetrant. The polymer chains in the rubbery state, being more mobile than in the glassy state, allow the active agent to diffuse out of the matrix more rapidly. The release rate is determined by the glassy-to-rubbery transition process (Buonocore et al. 2003a, b; Flores et al. 2007).

Among the abundant information dealing with release packaging systems, the current chapter focuses on eco-friendly active solutions obtained by the simple dispersion of antimicrobial/antioxidant compounds in the polymeric matrix and by the development of multilayer films.

5.2 Biopolymeric Active Systems by Direct Incorporation of Compounds

Today there is a growing interest from organizations, governments, and companies around the world to give sustainable development useful and practical meanings. In particular, sustainable packaging is a target vision for creating "a world where all packaging is sourced responsibly, designed to be effective and safe throughout its life cycle, meets market criteria for performance and cost, made entirely using renewable energy, and once used is recycled efficiently to provide a valuable resource for subsequent generations" (Lee et al. 2008). Actually, there are six ways to manage packaging wastes, usually combined to be more effective. The first three (reduce, reuse, and recycle) are aimed at minimizing the existence of postconsumer packaging materials. The remaining three ways (composting, incineration, and landfill) are aimed at disposing packaging wastes in an efficient and eco-friendly manner (Marsh and Bugusu 2007). In this context, numerous efforts have been made by researchers to develop active systems intended for food packaging applications based on renewable polymeric materials containing natural compounds with antimicrobial or antioxidant properties. A variety of bio-based materials have been shown to prevent moisture loss, reduce lipid oxidation, improve flavor, retain color, and stabilize the microbial characteristics of foods (Cutter 2006). Antimicrobial substances are defined as biocidal products under EU directives but would be permitted in food packaging only if there were no direct impact on the packaged-food quality. This requires that agent migration into food must not allow for the selection of biocide resistance in the microorganisms (McMillin 2008). Organic-based materials may be anaerobically degraded, while biodegradable polymers from agricultural feedstocks, animal sources, marine food processing industry wastes, or microbial sources are being developed (Marsh and Bugusu 2007). Biopolymer films may be potential replacements for synthetic films in food packaging applications to address strong marketing trends toward more environmentally friendly materials, but hydrophilicity is a central limitation on the replacement and full-scale commercial utilization of biodegradable films (McMillin 2008). Polysaccharides, proteins, and lipids are natural sources with film-forming properties that present numerous advantages such as biodegradability, edibility, biocompatibility, aesthetically pleasing appearance, and barrier properties against oxygen and physical stress. These polymers and their combinations are soluble in water, ethanol, and many other solvents compatible with antimicrobials (Ruban 2009; Tharanathan 2003). In particular, polysaccharides such as pectin, starch, cellulose, alginates, and other hydrocolloids have good performance in film forming due to their chemical nature. The addition of a plasticizer like glycerol or sorbitol increases the mobility of polymer chains because it reduces intermolecular forces, improving the flexibility and extensibility of the film (León et al. 2009). Bacterial cellulose is another source of bio-based packaging (Weber et al. 2002).

The poor water vapor resistance and their lower mechanical strength in comparison with synthetic polymers still limit the applications of bio-based systems as packaging films. Several studies have been carried out to improve the performance of protein films by incorporating hydrophobic compounds, by cross-linking polymer chains through chemical, enzymatic, or physical treatments, and, finally, by adding stiffening material in the film-forming solution to develop a biocomposite material (Mastromatteo et al. 2008, 2009b). The biocomposites consist of biodegradable polymers with biodegradable fillers, usually biofibers, such as cellulose fibers, cellulose whiskers, jute fibers, abaca fibers, pineapple fibers, flax fibers, wheat straw fibers or lignocellulosic flour, kenaf, spelt, and wheat-bran- or starch-based materials. High improvements in the performances (e.g., tensile and impact tests results) of biocomposite films have been found due to the effects of fiber addition that induces variation in film properties. Recently, film properties (rheological properties of film-forming solution and color, mass transport, and mechanical properties of films) of whey protein isolate (WPI) with different levels of glycerol and spelt bran were investigated using a response surface methodology to point out the individual and interactive effects of the selected variables (Mastromatteo et al. 2009b). Rheological investigation suggested that WPI solutions with a lower bran concentration behaved as a Newtonian fluid; as the bran concentration increased, the solutions demonstrated a non-Newtonian behavior. In addition, results on film properties highlighted that the water vapor permeability (WVP) was significantly affected by the spelt bran concentration. When the spelt bran concentration increased, the WVP decreased, whereas the converse happened with glycerol concentration. The elastic modulus (Ec) and complex modulus (E*) of the composite films increased with a decrease in the glycerol content and an increase in bran concentration.

Numerous studies have also been conducted from the late 1960s onward to improve the mechanical and physical properties of zein-based films by incorporating cross-linking agents, highly stable silicate complexes or fatty acids (Güçbilmez et al. 2007; Kim et al. 2004; Liu et al. 2005; Wang et al. 2003; Wu et al. 2003).

Relevant overviews were conducted to classify the numerous active bio-based systems, based on controlled or uncontrolled release; it is worth noting that most of them are devoted to describing applications to meat products (Aymerich et al. 2008; Coma 2008; Kerry et al. 2006; Mastromatteo et al. 2010; Quintavalla and Vicini 2002). The reason for this could be that the rationale for incorporating antimicrobials into packaging is the prevention of surface growth in foods where a large portion of spoilage and contamination occurs. In fact, intact meat from healthy animals is essentially sterile, and spoilage occurs primarily at the surface. The use of an active film can reduce the addition of larger quantities of antimicrobials that are usually spread on the food surface. The gradual release of an antimicrobial from a packaging film to the food surface may have an advantage over dipping and spraying. In the latter processes, antimicrobial activity may be rapidly lost due to inactivation of the antimicrobials by food components or dilution below the minimum active concentration as a result of migration to the bulk food matrix (Appendini and Hotchkiss 2002). Chitosan, alginate, and cellulose containing various bacteriocins, enzymes, essential oils, and organic acids are very diffuse examples of bio-based materials with antimicrobial and antioxidant properties. Lysozyme has been widely studied around the world (Fernandez et al. 2008). It is a lytic enzyme found in foods such as milk and eggs, and it is a muraminidase that hydrolyzes β1–4 linkages between N-acetylmuramic acid and N-acetylglucosamine. It is known to inhibit some Gram-positive bacteria, but alone it is ineffective against Gram-negative bacteria. The first example of physical retention of lysozyme was carried out by Appendini and Hotchkiss (1997) that entrapped the enzyme on polyvinyl alcohol (PVOH) beads, nylon 6.6 pellets, and cellulose triacetate (CTA) films, producing the best results on *M. lysodeikticus* when it was bound to CTA. Buonocore et al. (2003a, b) reported the suitability of highly swellable polymers to modulate the release kinetics of lysozyme and nisin by changing the degree of the chemically induced cross-linking. LaCoste et al. (2005) proposed the use of smart blending for developing novel controlled release packaging materials with active properties. Interesting results on the controlled release of lysozyme were also obtained by Mecitoglu et al. (2006), who worked on the feasibility of controlling the release rate of partially purified lysozyme incorporated into zein films containing salts of ethylenediaminetetraacetic acid (EDTA). The disadvantage of using partially purified lysozyme, which provoked a nonhomogenous distribution of the hydrophilic enzyme in hydrophobic zein films at high concentrations, was solved by the addition of edible ingredients to minimize structural changes and to obtain more uniform antimicrobial films. Hence, in a subsequent step the authors evaluated the efficacy of the partially purified lysozyme with EDTA incorporated into zein films, in combination with chickpea albumin extracts and bovine serum albumin against *Bacillus subtilis* and *Escherichia coli*, demonstrating the benefits of using functional protein extracts to control the lysozyme release rate and, consequently, its activity (Güçbilmez et al. 2007). The authors reported that Gram-negative bacteria are protected due to their outer membrane surrounding the peptidoglycan. Indeed, lysozyme exhibits antimicrobial activity by splitting the bonds between N-acetylmuramic acid and N-acetylglucosamine of the peptidoglycan in the cell wall of bacteria. The destabilization of the outer membrane by EDTA, potentially used in combination, could increase the activity on Gram-negative bacteria.

Another recent example of enzyme entrapment was given by Del Nobile et al. (2009a). These authors developed active films by extrusion to incorporate various polymeric matrix concentrations of thymol, lemon extract, and lysozyme, respectively. For the purposes of the work, low-density polyethylene (LDPE), polylactic acid (PLA), and polycaprolactone (PCL) were used as environmentally friendly polymeric matrices, the first one being recyclable and the latter two biodegradable. PLA is a degradable aliphatic polyester that can be produced synthetically or from renewable corn or whey resources and has food packaging applications thanks to its mechanical stability in ambient to chilled temperatures (Conn et al. 1995). PCL, a linear-polyester manufactured by ring-opening polymerization of ε-caprolactone, with a rather low glass transition temperature and melting point (65 °C), is an example of a petroleum-based biodegradable polymer (An et al. 2001). Del Nobile et al. (2009a) found that processing temperatures played a major role in determining the antimicrobial efficiency of active substrates. In particular, antimicrobials incorporated into PLA and LDPE retained slight antimicrobial effects, having lost some activity due to the high processing temperature. On the other hand, PCL, processed at a lower temperature, allowed less degradation of compound efficacy. Among the natural agents studied, lysozyme showed a higher thermal stability. From an industrial point of view, active films obtained using classical polymer technological processes, such as extrusion, are generally preferred. However, despite the great interest, few works have been reported (Nam et al. 2007), probably due to the deterioration that can occur during extrusion using high temperature, high shear rates, and, thus, high pressure. In fact, theoretically, thermal polymer processing methods, such as extrusion and injection molding, may be used with thermally stable antimicrobials, such as silver-substituted zeolites. For heat-sensitive antimicrobials like volatile compounds, solvent compounding may be a more suitable method for their incorporation into polymers. In solvent compounding, both the antimicrobial and the polymer need to be soluble in the same solvent; therefore, biopolymers are good candidates for this type of film-forming process (Appendini and Hotchkiss 2002).

A category of compounds that is attracting great interest in the field of food packaging is that of metallic nanoparticles, such as copper, zinc, titanium, magnesium, gold, and silver. The attention being given to the use of eco-friendly polymeric materials from natural sources or produced by synthesis, characterized by packaging with a low encumbrance, a low weight/volume ratio, and, where possible, a low cost, obliges us to solve the limitations of their scarce thermal, barrier, and mechanical properties. These materials, loaded with nanometric particles, appear extremely interesting as possible solutions to these problems. Moreover, it is possible to modify conveniently nanometric particles to make them *active*, thereby obtaining new materials with additional properties to be used for prolonging the SL of packaged commodities. Among the available nanoparticles, silver nanoparticles have attracted significant interest due to their effectiveness against a wide range of microorganisms. For this reason silver has been used as either an additive or coating in/on various polymers to demonstrate antimicrobial properties (An et al. 2008; Del Nobile et al. 2004; Fernandez et al. 2009). The mechanism of the antimicrobial activity of silver is well known: in the presence of water and oxygen, elemental silver particles release small amounts of silver ions, according to the following reaction:

$$O_{2\,(aq)} + 4H_3O^+ + 4Ag_{(s)} \rightarrow 4Ag^+_{\,(aq)} + 6H_2O.$$

Silver ions form complexes with sulfur-, nitrogen-, and oxygen-containing functional groups of organic compounds present in bacteria. This may result in defects in cell walls so that plasma is lost. A complex formation between silver ions and proteins may disturb the metabolism of the bacterial cells. Both effects lead to death of bacterial cells (bactericidal activity of silver). Moreover, silver ions can interact with the DNA of bacteria, preventing cell reproduction (bacteriostatic activity of silver). Inorganic phyllosilicate clays have been used as support for silver nanoparticles to generate a new class of antimicrobial systems from which silver ions can be released into media (Choudalakis and Gotsis 2009; Dong et al. 2009). In this context, Incoronato et al. (2010) obtained Ag^+-montmorillonite (Ag-MMT) nanoparticles by allowing silver ions from silver nitrate ($AgNO_3$) to replace the exchangeable Na^+ counterions in the natural sodium montmorillonite and without using any reducing agent. Bearing in mind that the release of Ag^+ ions into a medium may be ruled by weak electrostatic interactions that are established through Ag ions and surface platelets of MMT, the release kinetics of silver ions could be delayed. Antimicrobial nanocomposites have been obtained by embedding various amounts of Ag^+-MMT in agar, zein, and poly(ε-caprolactone) (PCL) as biopolymer matrices. In vitro tests were carried out on a microbial cocktail of three strains of *Pseudomonas* spp. to prove their antimicrobial effectiveness. UV-visible analyses confirmed that silver ions were not reduced to metal silver, whereas X-ray diffraction analysis proved that MMT clays lose their layered structure, giving rise to nanoparticles with a collapsed or exfoliated structure. The results related to antimicrobial efficacy showed that the hydration level of the organic matrix played a key role in exerting a good antimicrobial effect. In fact, due to the circumstance that agar hydrogel contained the highest water content, agar-based nanocomposites presented the highest activity against the selected microorganisms. As can be observed in Fig. 5.1, the cell loads attained at the stationary phase are much lower for the active systems based on agar, compared to the respective control samples. No antimicrobial effects were recorded with active zein and PCL, probably due to the different hydrophilicity of the polymer matrix compared to the polysaccharide-based layer. In fact, previous studies on silver-based antimicrobial nanofiller embedded into polymer matrices determined that the amount of the released active species (Ag^+) strongly depends on the hydrophilic nature of the polymeric matrix and on the boundary conditions of the outer solution, such as pH and ionic strength. Of the outer solution (Klasen 2000; Yoksan and Chirachanchai 2009).

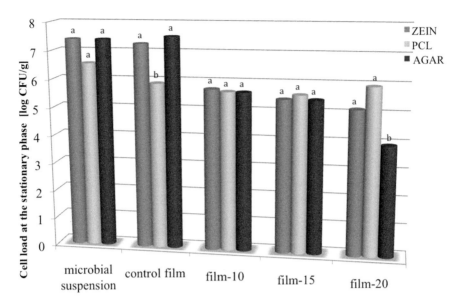

Fig. 5.1 Cell load attained at stationary phase in microbial suspension with no film, with a control film, and with three active systems releasing silver ions. The value of the cell load is calculated by fitting the Gompertz equation to the experimental data. Statistically significant differences ($p < 0.05$) between samples of each group are identified by different letters. *Control film* system with film free of nanoparticles, *Film*-10 system with 10 mg of silver-montmorillonite nanoparticles, *Film*-15 system with 15 mg of silver-montmorillonite nanoparticles, *Film*-20 system with 20 mg of silver-montmorillonite nanoparticles

An alternative to incorporating antimicrobial compounds by extrusion or solvent compounding is to apply an additive as a coating (Appendini and Hotchkiss 2002; Chung et al. 2003). This has the advantage of placing specific antimicrobials in a controlled manner without subjecting them to high temperature or shearing forces. In addition, the coating can be applied at a later step, minimizing the exposure of the product to contamination (Coma 2008). The coating can serve as a carrier for antimicrobial compounds in order to maintain high concentrations of preservatives on the surface of foods. Bioactive activity may be based on migration or release by evaporation in the headspace (Lee et al. 2004; Skandamis and Nychas 2002). A valid polymeric structure with a significant controlled release of lysozyme, intended for internal food packaging applications, was represented by an asymmetric porous cellulose acetate-based film containing lysozyme, developed by Gemili et al. (2009). Other works showed the advantage of using asymmetric-membrane capsules and asymmetric coatings on drug tablets to control the release rate of drugs (Altinkaya and Yenal 2006). However, the work of Gemili et al. (2009) represents the first tentative step toward clearly demonstrating that porous films can be used as novel internal food packaging systems with controlled release properties and could be tailored by changing the composition of the initial casting solution. Further studies are still needed to test the effectiveness of these active structures on real food systems.

Another possibility for realizing bio-based systems with active properties is to incorporate the antimicrobial compound into an edible coating, directly applied by dipping or spraying onto the food surface (Conte et al. 2009; Del Nobile et al. 2009b, c; Coma 2008). Edible coatings applied on food surfaces could help to alleviate the problem of moisture loss during storage; separate incompatible zones and ingredients of a food matrix; form a barrier against oxygen, aroma, and oil; hold the juices of fresh-cut food; reduce the rate of rancidity; reduce nonenzymatic browning; reduce the load of spoilage and pathogenic microorganisms on food surfaces; and restrict volatile flavor loss and foreign odor pickup (Appendini and Hotchkiss 2002; Cutter 2006; Coma 2008). The selection of the incorporated active agents is limited to edible compounds because they are consumed with edible film/coating layers and food together. A large number of scientific articles and reviews dealing with the antimicrobial activity of edible coatings is available in the literature (McMillin 2008; Aymerich et al. 2008; Coma 2008, Hernandez-Izquierdo and Krochta 2008; Matromatteo et al. 2010). However, it is difficult to compare the results of these studies because of the substantial variations in bioactive compounds, polymeric materials, test microorganisms, and test methods. Among the bio-based materials, chitosan represents a particular case due to both antimicrobial activity and film-forming properties. Chitosan consists of polymer composed principally of 1,4 linked 2-amino-2-deoxy-b-D-glucose and, although more active against spoilage yeasts, it also inhibits some Gram-negative and particularly Gram-positive bacteria. It has been affirmed as GRAS by the U.S. FDA, thus removing some of the regulatory restrictions on its use in foods. Due to its nature, it can be used as an active compound or as carrier of other natural preservatives (Devlieghere et al. 2004; No et al. 2007; Wang et al. 2007). For example, Pranoto et al. (2005) showed that the incorporation of garlic oil into chitosan film improved the antimicrobial efficacy of chitosan against food pathogenic bacteria. Vanillin incorporated into chitosan/methyl cellulose films provided an inhibitory effect against *E. coli* and *Saccharomyces cerevisiae* on fresh-cut cantaloupe and pineapple stored at 10 °C (Sangsuwan et al. 2008). Ponce et al. (2008) found that the use of chitosan coatings enriched with oleoresins applied to butternut squash did not produce a significant antimicrobial effect. However, it improved the antioxidant protection of minimally processed squash, offering a great advantage in the prevention of browning reactions that typically result in quality loss in fruits and vegetables. Some interesting applications of chitosan as a coating for fruits and vegetables have been also cited in the review of Mastromatteo et al. (2009c) dealing with preservation techniques of fresh-cut produce.

Modeling bioactive substance diffusion is crucial for understanding or modulating film activity and for investigating what type of food could be protected efficiently using these systems of active films. Several approaches have been reported in the literature that quantitatively take into account the stochastic diffusion (related to Brownian motion) and the relaxation phenomenon involved in mass diffusion (Del Nobile et al. 1996). To this end, Buonocore et al. (2003a, b) proposed a complex approach that can predict the compound release kinetic from cross-linked PVOH into aqueous solution. The model was developed taking into account the diffusion of water molecules into a polymeric film and the counterdiffusion of the incorporated antimicrobial agent from the film into the aqueous solution, also considering the dependence of the water diffusion coefficient on the degree of cross-link. Therefore, two models, one describing the water uptake kinetic and the other describing the enzyme release, were taken into account (Crank 1955). The first model was fitted to the water sorption data, and the obtained parameters were used in the second model to fit the release data. The mass balance for the penetrant and the active substance was numerically solved by means of the finite-element method. Flores et al. (2007) proposed a simple mathematical model that can describe the release of sorbate from tapioca starch edible film. The authors suggested that, despite the complexity of the phenomena involved during the release of an active compound from a polymeric matrix, the adopted model can be used to obtain useful information on the release mechanism of low molecular weight compounds from films. The model was based on the assumption that a linear superimposition of stochastic diffusion and relaxation phenomenon occurs, thereby combining both mechanisms. Mass transport related to Brownian motion was described by means of Fick's second law (Langer and Peppas 1983), while polymer relaxation was driven by the distance of the system from equilibrium and was quantitatively described through a first-order kinetic-type equation. More recently, Del Nobile et al. (2008) studied the antimicrobial properties of a controlled release system based on a zein film containing different concentrations of thymol. The release kinetic of the selected preservative from the biopolymer to a liquid medium was also monitored to investigate the dependence of the diffusion coefficient on the loaded agent concentration. Also in this case the swelling phenomenon was taken into account. Thus, the two mechanisms considered were the penetration of water molecules into the matrix and the active compound diffusion through the matrix into the outer solution until reaching a thermodynamic equilibrium between the two phases. According to the approach previously described by Buonocore et al. (2003a, b), both water diffusion and macromolecular matrix relaxation were assumed to be much faster than the active compound diffusion through the swollen network. Therefore, the active compound release kinetic was described by means of Fick's second law, intended for a plane sheet with constant boundary condition and uniform initial concentration. The equation was also reported in Chap. 2; for the sake of simplicity it is also reported below:

$$M_{Thy}(t) = M_{Thy}^{eq.} \cdot \left\{ 1 - \frac{8}{\pi^2} \cdot \sum_{n=0}^{n=\infty} \frac{1}{(2 \cdot n + 1)^2} \cdot \exp\left\{ -D \cdot (2 \cdot n + 1)^2 \cdot \pi^2 \cdot \frac{t}{\ell^2} \right\} \right\},$$

(5.1)

where $M_{Thy}(t)$ is the amount of thymol released at time t, $M_{Thy}^{eq.}$ is the amount of thymol released at equilibrium, D is the thymol diffusion coefficient through the swollen polymeric matrix, and ℓ is the film thickness. The thymol diffusion coefficient slightly increased with thymol concentration (Fig. 5.2). From the figure it is

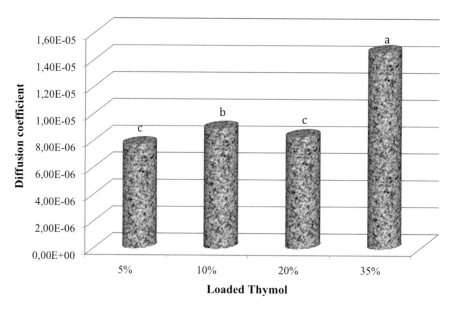

Fig. 5.2 Thymol diffusion coefficient through swollen zein polymeric matrix, calculated as fitting parameter. Statistically significant differences ($p < 0.05$) between samples are underlined by different letters

possible to infer that the release kinetic does not depend much on the thymol concentration. In addition, the results obtained from the released thymol were compared to those obtained by direct addition of the active compound to a *Pseudomonas* microbial suspension. Either when thymol was released from the film or when it was directly added to the media, a complete microbial growth inhibition was observed. However, considering that essential oils could impart off-odors to the food, their incorporation in a polymeric matrix could be a strategic solution to exert the desired antimicrobial function without altering the sensory characteristics of the food (Suppakul et al. 2003a).

5.3   Multilayer Films to Control Active Agent Release

A multilayer system represents a solution to achieve a controlled release of active compounds from packaging system to food surface (Floros et al. 2000). A multi-layer design has the advantage that the antimicrobial can be added in one thin layer and its migration is controlled by the thickness of the film layer or coating. As was reported earlier, the control of the release rate and amount of antimicrobial substances is very important. A mass transfer model of the migration phenomenon can be used to describe the concentration profile in the film/coating layer and food over the storage time. Han (2000) summarized traditional mass transfer systems and proposed models that may be used to describe the migration of active agents through food packaging systems consisting of single, double, or triple layers. Using mass transfer models it is possible to calculate the storage periods to maintain the active agent concentration above the critical effectiveness concentration, allowing for the calculation of the food's safety SL.

Moreover, it could be considered that multilayer systems represent a need for recycled plastics. Recycling is becoming more common, requiring functional barriers in multilayer structures to be used between the recycled plastic and food to prevent food contact by any contaminants in the recycled plastics (Feigenbaum et al. 2005).

The system can be constituted by control layer/active matrix layer/barrier layer. The outer layer is the barrier to prevent loss of active substances to the environment, the matrix layer contains an active substance and has a very fast diffusion, and the control layer is the key layer to control the flux of penetration; it has a tailored thickness and diffusivity with respect to the characteristics of microbial spoilage of food products.

Taking into account the preceding structure, Buonocore et al. (2005) developed two multilayer films, both composed of two external control layers and an inner layer containing lysozyme. A comparison between the release rate attained with these multilayer films and by a monolayer cross-linked PVOH film studied in a previous work (Buonocore et al. 2003a) was also carried out. Results showed that by means of the multilayer structure it is possible to control the rate at which lysozyme is released from a swellable PVOH film; comparing the behavior of monolayer films cross-linked with glyoxal with multilayer films, it was pointed out that the latter system was more effective in slowing down the lysozyme release rate. Moreover, the findings highlighted that the amount of lysozyme released at equilibrium was strongly influenced by the total amount of cross-linking agent and by the layer that coincided with the active layer in the monolayer structure and with the external layers in the multilayer system. As regards the antimicrobial effectiveness of the released lysozyme, similar effects were recorded with the active films on the microbial target microorganism. The observed decrease in optical density shows that the incorporation of lysozyme into PVOH did not lead to a loss of activity of the enzyme, regardless of the technique used to control its release from the polymeric film.

More recently, Mastromatteo et al. (2009a) developed zein-based mono- and multilayer films loaded with spelt bran and thymol (35 % w/w) to obtain edible composite materials to control thymol release. To this end, the thickness of the film layers and the amount of biodegradable fibers were varied. Considering the difficulties in modeling thymol release, due to the numerous phenomena involved, a very simple approach was used assuming that (1) both water diffusion and macromolecular matrix relaxation are faster than active compound diffusion through a swollen network, (2) the increase in the film size due to swelling is negligible, and (3) thymol diffusion takes place in an homogeneous and symmetric medium. As a consequence, the active compound release kinetic was described by means of Fick's second law (Eq. 5.1). Because of the numerous hypotheses made

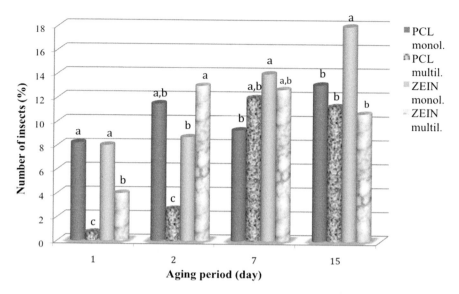

Fig. 5.3 Percentage of *Sitophilus granarius* that entered cartons with different coatings activated with propionic acid. Statistically significant differences ($p < 0.05$) between samples of each group are identified by different letters. *PCL monol.* monolayer film based on PCL, *PCL-multil.* multilayer film based on PCL, *Zein monol.* monolayer film based on zein, *ZEIN-multil.* multilayer film based on zein

for deriving Eq. 5.1, the value of D_{Thy} must be considered as an apparent diffusion coefficient and not as the thymol thermodynamic diffusion coefficient. For this reason, the ratio D_{Thy}/ℓ^2 was determined in place of D_{Thy}, with the advantage that the new term was directly related to the rate at which thymol was released from the film into the outer water solution. Results highlighted that the release rate decreased with the increase in film thickness for both mono- and multilayer films, without the addition of spelt bran. Conversely, a significant increase in the thymol release rate with an increase in bran concentration was recorded for both mono- and multilayer films, probably due to the fact that the addition of spelt bran promoted the creation of the microchannels interconnected with the thymol phase, bypassing the zein matrix and leading to an increase in the active compound release rate.

An interesting work dealing with the effectiveness of multilayer systems was presented by Germinara et al. (2010). The authors developed a biodegradable carrier material to control insect pests in cereal products. To that end, propionic acid, a known repellent of stored-product insects, was incorporated into biodegradable coatings made up of different layers of corn zein and policaprolactone (PCL), respectively. The bio-based multilayer coatings were applied to cartons intended for packaging cereal goods. To simulate extreme infestation conditions, the bioassay was carried out over a 2-week period with a high number of insects confined in a small space and damaged cartons imitating breakdown or poorly sealed packaging. As can be seen in Fig. 5.3, at the end of the aging period, the percentage of insects found in cartons coated with propionic-acid-loaded mono- and multilayer PCL and zein was only 13.1, 11.3, 18.0, and 10.7 % of the total number of insects used in the bioassay, respectively. In contrast, coating materials or solvents used for coating preparation had no effect on the insects' ability to enter the cartons since the percentages of insects found in coated cartons without propionic acid and in the uncoated cartons were almost identical.

References

Altinkaya SA, Yenal H (2006) In vitro drug release rates from asymmetric- membrane tablet coatings: prediction of phase-inversion dynamics. Biochem Eng J 28:131–139

An JH, Kim HS, Chung DJ, Lee DS (2001) Thermal behaviour of poly(ecaprolactone)-poly (ethylene glycol)-poly(e-caprolactone) tri-block copolymers. J Mat Sci 36:715–722

An J, Zhang M, Wang S, Tang J (2008) Physical, chemical and microbiological changes in stored green asparagus spears as affected by coating of silver nanoparticles-PVP. LWT Food Sci Technol 41:1100–1107

Appendini P, Hotchkiss JH (1997) Immobilization of lysozyme on food contact polymers as a potential antimicrobial films. Packag Technol Sci 10:271–279

Appendini P, Hotchkiss JH (2002) Review of antimicrobial food packaging. Innov Food Sci Emerg Technol 3:113–126

Aymerich T, Picouet PA, Monfort JM (2008) Decontamination technologies for meat products. Meat Sci 78:114–129

Buonocore GG, Del Nobile MA, Panizza A, Bove S, Battaglia G, Nicolais L (2003a) Modeling the lysozyme release kinetics from antimicrobial films intended for food packaging applications. J Food Sci 68:1365–1370

Buonocore GG, Del Nobile MA, Panizza A, Corbo MR, Nicolais L (2003b) A general approach to describe the antimicrobial agent release from highly swellable films intended for food packaging applications. J Control Release 90:97–107

Buonocore G, Conte A, Corbo MR, Sinigaglia M, Del Nobile MA (2005) Mono- and multilayer active films containing lysozyme as antimicrobial agent. Innov Food Sci Emerg Technol 6:459–464

Choudalakis G, Gotsis AD (2009) Permeability of polymer/clay nanocomposites: a review. Eur Polym J 45:967–984

Chung D, Papadakis SE, Yam KL (2003) Evaluation of a polymer coating containing triclosan as the antimicrobial layer for packaging materials. Int J Food Sci Technol 38:165–169

Coma V (2008) Bioactive packaging technologies for extended shelf life of meat-based products. Meat Sci 78:90–103

Conte A, Scrocco C, Brescia I, Del Nobile MA (2009) Packaging strategies to prolong the shelf life of minimally processed lampascioni (Muscari comosum). J Food Eng 90:199–206

Conn RE, Kolstad JJ, Borzelleca JF, Dixler DS (1995) Safety assessment of polylactide (PLA) for use as a food-contact polymer. Food Chem Toxicol 33:273–283

Cooney DO (1972) Effect of geometry on the dissolution of pharmaceutical tablets and other solids: surface detachment kinetics controlling. AIChE J 18:446–449

Crank J (1955) The mathematics of diffusion. Clarendon Press, Oxford, pp 56–60

Cutter CN (2006) Opportunities for bio-based packaging technologies to improve the quality and safety of fresh and further processed muscle foods. Meat Sci 74:131–142

Del Nobile MA, Mensitieri G, Manfredi C, Arpaia A, Nicolais L (1996) Low molecular weight molecules diffusion in advanced polymers for food packaging applications. Polym Adv Technol 7:409–417

Del Nobile MA, Cannarsi M, Altieri C, Sinigaglia M, Favia P, Iacoviello G, D'Agostino R (2004) Effect of Ag-containing nano-composite active packaging system on survival of Alicyclobacillus acidoterrestris. J Food Sci 69:379–383

Del Nobile MA, Conte A, Incoronato AL, Panza O (2008) Antimicrobial efficacy and release kinetics of thymol from zein films. J Food Eng 89:57–63

Del Nobile MA, Conte A, Buonocore GG, Incoronato AL, Massaro A, Panza O (2009a) Active packaging by extrusion processing of recyclable and biodegradable polymers. J Food Eng 93:1–6

Del Nobile MA, Conte A, Scrocco C, Laverse J, Brescia I, Conversa G, Elia A (2009b) New packaging strategies to preserve fresh-cut artichoke quality during refrigerated storage. Innov Food Sci Emerg Technol 10:128–133

Del Nobile MA, Conte A, Scrocco C, Brescia I (2009c) New strategies for minimally processed cactus pear packaging. Innov Food Sci Emerg Technol 10:356–362

Devlieghere F, Vermeulen A, Debevere J (2004) Chitosan: antimicrobial activity, interactions with food components and applicability as a coating on fruit and vegetables. Food Microbiol 21:703–714

Dong RX, Chou CC, Lin JJ (2009) Synthesis of immobilized silver nanoparticles on ionic silicate clay and observed low-temperature melting. J Mater Chem 19:2184–2188

Feigenbaum A, Dole P, Aucejo S, Dainelli D, De La Cruz Garcia C, Hankemeier T et al (2005) Functional barriers: properties and evaluation. Food Addit Contam 22:980–987

Fernandez A, Cava D, Ocio MJ, Lagaron JM (2008) Perspective for biocatalysts in food packaging. Trends Food Sci Technol 19:198–206

Fernandez A, Soriano E, Lopez G, Carballo P, Picouet E, Lloret E, Gavara R, Hernandez-Munoz P (2009) Preservation of aseptic conditions in absorbent pads by using silver nanotechnology. Food Res Int 42:1105–1112

Flores S, Conte A, Campos C, Gerschenson L, Del Nobile MA (2007) Mass transport properties of tapioca-based active edible films. J Food Eng 81:580–586

Floros J, Nielsen P, Farkas J (2000) Advances in modified atmosphere and active packaging with applications in the dairy industries. Bull Int Dairy Fed 346:22–28

Gemili S, Yemenicioğlu A, Altınkaya SA (2009) Development of cellulose acetate based antimicrobial food packaging materials for controlled release of lysozyme. J Food Eng 90:453–462

Germinara SG, Conte A, Lecce L, Di Palma A, Del Nobile MA (2010) Propionic acid in bio-based packaging to prevent *Sitophilus granarius* (L.) (Coleoptera, Dryophthoridae) infestation in cereal products. Innov Food Sci Emerg Technol 11:498–502

Güçbilmez ÇM, Yemenicioğlu A, Arslanoğlu A (2007) Antimicrobial and antioxidant activity of edible zein films incorporated with lysozyme, albumin proteins and disodium EDTA. Food Res Int 40:80–91

Han JH (2000) Antimicrobial food packaging. Food Technol 54:56–65

Hernandez-Izquierdo VM, Krochta JM (2008) Thermoplastic processing of proteins for film formation: a review. J Food Sci 73:30–39

Incoronato AL, Buonocore GG, Conte A, Lavorgna M, Del Nobile MA (2010) Active systems based on silver/montmorillonite nanoparticles embedded into bio-based polymer matrices for packaging applications. J Food Prot 73:2256–2262

Kerry JP, O'Grady MN, Hogan SA (2006) Past, current and potential utilization of active and intelligent packaging systems for meat and muscle-based products: a review. Meat Sci 74:113–130

Kim S, Sessa DJ, Lawton JW (2004) Characterization of zein modified with a mild cross-linking agent. Ind Crops Prod 20:291–300

Klasen HJ (2000) A historical review of the use of silver in the treatment of burns. Part I early uses. Burns J 30:1–9

LaCoste A, Schaich KM, Zumbrunnen D, Yam KL (2005) Advancing controlled release packaging through smart blending. Packag Technol Sci 18:77–87

Langer R, Peppas NA (1983) Chemical and physical structure of polymers as carriers for controlled release of bioactive agents: a review. Macromol Chem Phys 23:61–126

Lee CH, An DS, Lee SC, Park HJ, Lee DS (2004) A coating for use an antimicrobial and antioxidative packaging material incorporating nisin and α-tocopherol. J Food Eng 62:323–329

Lee DS, Yam KL, Piergiovanni L (2008) Food packaging science and technology. CRC Press, Boca Raton, pp 595–597

León P, Chillo S, Conte A, Gerschenson L, Del Nobile MA, Rojas AM (2009) Rheological characterization of deacylated/acylated-gellan films carrying L-(+)-ascorbic acid. Food Hydrocolloid 23:1660–1669

Liu X, Sun Q, Wang H, Zhanga L, Wang JL (2005) Microspheres of corn protein, zein, for an ivermectin drug delivery system. Biomaterials 26:109–115

Marsh K, Bugusu B (2007) Food packaging – roles, materials, and environmental issues. Food Sci 72:39–55

Mastromatteo M, Chillo S, Buonocore GG, Massaro A, Conte A, Del Nobile MA (2008) Effects of spelt and wheat bran on the performances of wheat gluten films. J Food Eng 88:202–212

Mastromatteo M, Barbuzzi G, Conte A, Del Nobile MA (2009a) Controlled release of thymol from zein based film. Innov Food Sci Emerg Technol 10:222–227

Mastromatteo M, Chillo S, Buonoocre GG, Massaro A, Conte A, Bevilacqua A, Del Nobile MA (2009b) Influence of spelt bran on the physical properties of WPI composite films. J Food Eng 92:467–473

Mastromatteo M, Conte A, Del Nobile MA (2009c) Preservation of fresh-cut produce using natural compounds. A Rev Stewart Postharvest Rev 4:4

Mastromatteo M, Mastromatteo M, Conte A, Del Nobile MA (2010) Advances in controlled release devices for food packaging applications. Trends in Food Sci Technol 21:591–598

McMillin KV (2008) Where is MAP going? A review and future potential of modified atmosphere packaging for meat. Meat Sci 80:43–65

Mecitoglu C, Yemenicioglu A, Arslanoglu A, Elmaci ZS, Korel F, Cetin AE (2006) Incorporation of partially purified hen egg white lysozyme into zein films for antimcrobial food packaging. Food Res Int 39:12–21

Nam S, Scanlon MG, Han JH, Izydorczyk MS (2007) Extrusion of pea starch containing lysozyme and determination of antimicrobial activity. J Food Sci 72:477–484

No HK, Meyers SP, Prinyawiwatkul W, Xu Z (2007) Application of chitosan for improvement of quality and shelf life of foods – a review. J Food Sci 72:87–100

Ponce AG, Roura SI, del Valle CE, Moreira MR (2008) Antimicrobial and antioxidant activities of edible coatings enriched with natural plant extracts: in vitro and in vivo studies. Postharvest Biol Technol 49:294–300

Pothakamury UR, Barbosa-Cánovas GV (1995) Fundamental aspects of controlled release in foods. Trends Food Sci Technol 6:397–406

Pranoto Y, Rakshit SK, Salokhe VM (2005) Enhancing antimicrobial activity of chitosan films by incorporating garlic oil, potassium sorbate and nisin. Lwt-Food Sci Technol 38:859–865

Quintavalla S, Vicini L (2002) Antimicrobial food packaging in meat industry. Meat Sci 62:373–380

Ruban SW (2009) Biobased packaging-application in meat industry. Vet World 2:79–82

Sangsuwan J, Rattanapanone N, Rachtanapun P (2008) Effect of chitosan/methyl cellulose films on microbial and quality characteristics of freshcut cantaloupe and pineapple. Postharvest Biol Technol 49:403–410

Skandamis PN, Nychas GJE (2002) Preservation of fresh meat with active and modified atmosphere packaging conditions. Int J Food Microbiol 79:35–45

Suppakul P, Miltz J, Sonneveld K, Bigger SW (2003a) Active packaging technologies with an emphasis on antimicrobial packaging and its applications. J Food Sci 68:408–420

Suppakul P, Miltz J, Sonneveld K, Bigger SW (2003b) Antimicrobial properties of basil and its possible application in food packaging. J Agric Food Chem 51:3197–3207

Tharanathan RN (2003) Biodegradable films and composite coatings: past, present and future. Trends Food Sci Technol 14:71–78

Wang Q, Crofts AR, Padua GW (2003) Protein-lipid interactions in zein films investigated by surface plasmon resonance. J Agric Food Chem 51:7439–7444

Wang X, Du Y, Luo J, Lin B, Kennedy JF (2007) Chitosan/organic rectorite nanocomposite films: structure, characteristic and drug delivery behaviour. Carbohydr Polym 69:41–49

Weber CJ, Haugaard V, Festersen R, Bertelsen G (2002) Production and applications of biobased packaging materials for the food industry. Food Addit Contam 19(Suppl):172–177

Wu Q, Yoshino T, Sakabe H, Zhang H, Isobe S (2003) Chemical modification of zein by bifunctional polycaprolactone (PCL). Polymer 44:3909–3919

Yoksan R, Chirachanchai S (2009) Silver nanoparticles dispering in chitosan solution: preparation by γ-ray irradiation and their antimicrobial activities. Mater Chem Phys 115:296–302

Part III
New Strategies to Prolong Food Shelf Life

Product shelf life is defined as the period of time during which the quality of the packaged food remains acceptable. This period may range from a few days to over a year, depending on product characteristics, food processing, packaging, and storage conditions. Conventional food preservation methods often result in a number of undesired changes in foods, such as loss of smell, color, flavor, texture, and nutritional value. In addition, consumers from developing countries are more concerned with the nutritional and sensory aspects, as well as the safety, of the food they eat. Therefore, a more differentiated food product assortment and novel preservation strategies are necessary for today's increasingly demanding consumers. Considering the importance of packaging in determining product shelf life, the correct approach allows considering on the same level of importance product development and packaging system. Food is packaged to preserve its quality and freshness. Packaging acts as a physical barrier to gas, moisture, external compounds, and microorganisms that could be detrimental to food. The preservation role is a fundamental requirement of food packaging since it is directly related to consumer safety. Numerous variables play a significant role in establishing package performance, such as the initial food quality, processing operations, size, shape and appearance of the package, distribution method, and package disposal. Generally speaking, packaging properties can be grouped into mechanical, thermal, optical, and mass transport properties, but the extent to which packaging plays a preservation role is largely dependent on the barrier of the material to environmental factors that cause spoilage. There are occasions when the transport of gases is desirable, as happens in fresh-cut produce where the exchange of gas through the package is necessary to accommodate respiration and transpiration, and cases where high-barrier properties are preferred. Similarly, if the package is too permeable to water vapor, it causes a moisture-sensitive food to have less crispness and, consequently, a shorter shelf life; on the other hand, water vapor escaping from the package can provoke undesirable textural changes in wet food. For a specific shelf life package characteristics can be determined from the interactions between food, packaging, and environment. The current section aims to review a number of case studies related to various food categories where the proper selection of packaging conditions may greatly prevent product deterioration, thereby promoting a significant shelf life prolongation. Liquid foods, minimally processed products, dairy food, and meat- and fish-based goods will be taken into account as main food categories. The main spoilage changes that affect these products, as well as the traditional processing and preservation techniques, are reviewed. Additionally, the various chapters focus on the keys to the production of safe foods and, in particular, on some successful combinations of inhibitory processes based on the application of various mild treatments that take advantage of the synergisms among the different preservation hurdle technologies.

Chapter 6
New Packaging for Food Beverage Applications

6.1 Introduction

Liquid foods include different types of products, fruit juices, soft drinks, beers, wines, milk, water, oils, etc. Each category of food has its specificity in quality attributes, storage conditions, expected shelf life (SL), and packaging tools applied (Dalpasso 1991). The quality deterioration mode differs with food type, and the prevailing mode of a food may also be determined by the storage and packaging conditions. The protection provided by packages includes the proper evaluation of physical, chemical, and biological factors interacting with the package and the environment. Therefore, packaging requirements can be estimated by applying a SL model to the expected conditions (Robertson 1993). For example, a small plastic package size for oil may reduce the span of storage in use; however, it could increase the ratio of surface area to oil volume and enhance the rate of oxygen ingress by permeation through the package wall. The incorporation of oxygen scavengers in containers was found to be an interesting solution to protecting oil from oxidation (Del Nobile et al. 2003a), as assessed in detail in Chap. 3 of the modeling section. In the case of fruit juice, major quality deterioration modes are the destruction of ascorbic acid and carotenoids, nonenzymatic browning, and other oxidative reactions of flavors. Packages of glass, metal cans, plastic bottles, and paper cartons laminated with aluminum foil provide a proper barrier to oxygen, preventing ascorbic acid depletion. However, some differences exist between the various packaging systems. Canned juice of high acidity may suffer from the corrosion of tin; therefore, lacquered cans must be used to avoid this problem. In turn, ascorbic acid retention is higher for plain cans than lacquered cans because of oxygen consumption by tin dissolution. Glass and plastic bottles are transparent to light and thus contribute to juice deterioration. Moreover, for juice packaged in plastic materials, just as for other drinks in plastics, there is a concern about flavor adsorption on plastic walls, the so-called phenomenon of scalping that further degrades sensory quality (Nielsen and Jagerstad 1994; Licciardello et al. 2009). The most numerous studies concern flavor scalping by low-density poly(ethylene)

M.A. Del Nobile and A. Conte, *Packaging for Food Preservation*,
Food Engineering Series, DOI 10.1007/978-1-4614-7684-9_6,
© Springer Science+Business Media New York 2013

(LDPE) and particularly the sorption of limonene and other aroma compounds from fruit juice (Fayoux et al. 1997; Askar 1999). LDPE is frequently used for contact with beverages and has been proven to be very prone to sorption of such molecules.

Flavor degradation and loss of carbon dioxide are the two main concerns that affect the sensory characteristics of carbonated soft drinks. For improved protection of color and flavor, colored glass is generally used for soft drinks. In the case of plastic bottles an additional layer with an ultraviolet ray barrier may be employed. To minimize carbon dioxide loss, a small ratio of surface area to volume is advised. Del Nobile et al. (1997) demonstrated that temperature control is very important to attain the desired retention of carbon dioxide in packages. Beer is similar to carbonated soft drinks in the required properties for packaging, but it is more sensitive to flavor deterioration stimulated by oxygen and light, thus limiting packaging systems to barrier metal cans and glass bottles (Kuchel et al. 2006). Devices for easy opening and dispensing (snap-on caps, peelable lids, or tear-away strips) have also been recently emphasized for the convenience of drink usage.

This chapter analyzes the quality change aspects of various liquid foods and presents a method for designing the packaging. Due to the numerous differences between the various food categories and to the impossibility of dealing with all the food types in one chapter, two typical Apulian foods were chosen, olive oil and wine. In particular, the chapter focuses on the scientific progress aimed at designing proper packaging and presents an overview dealing with the most recent packaging systems intended for olive oil and wine applications.

6.2 Packaging of Olive Oil: Quality and Shelf Life Prediction

Olive oil is one of the most important and ancient oils in the world. It contains 56.3–86.5 % monounsaturated fatty acids, 8–25 % saturated fatty acids, and 3.6–21.5 % polyunsaturated fatty acids, thus justifying its reputation as being the most monounsaturated vegetable oil (IOOC 1984). Olive oil has a high biological value due to its high ratio of vitamin E to polyunsaturated fatty acid content and to the high content of phenolic substances that act as antioxidant compounds (Baldioli et al. 1996; Tsimidou et al. 1992; Viola 1970). It is the most widely used oil in those countries that border the Mediterranean Sea. As a matter of fact, the incidence of heart disease is relatively low in Mediterranean countries due to the diet rich in monounsaturated fats (olive oil) (Keys 1970; Keys et al. 1986). It is used almost entirely for edible purposes as a cooking and salad oil (Tawfik and Huyghebaert 1999).

To package vegetable oil, glass, metals, and various plastic packages are used with different effects on the product, depending on the barrier properties against moisture, oxygen, and interactions of food constituents with the packaging materials. Glass containers are generally preferred to plastic ones for bottling virgin olive oil, in part due to marketing considerations and in part due to the fact that glass containers prevent the permeation of oxygen into bottles, thereby slowing down the rate at which the oxidation reaction of unsaturated fatty acids proceeds in glass bottles compared to their plastic counterparts. It is well known that the SL of olive oil is limited by the oxidation of unsaturated fatty acids with the formation of hydroperoxides, which in turn give rise to a complex mixture of compounds responsible for oil degradation (Frankel 1998; Kanavouras et al. 2004; Mastrobattista 1990). The rate of oxidation depends mainly on storage conditions, such as temperature and presence of light, as well as on the availability of soluble and reactive oxygen in the oil's mass (Angelo 1996; Labuza 1971).

Due to the low weight, easier handling, and competitive costs, more effort has been made to replace glass containers with plastic materials (Franz et al. 1996; Kiritsakis et al. 2002). However, plastics offer limited protection regarding their gas barrier properties and migration of compounds compared to steel and glass (Riquet et al. 1998). Kiritsakis and Dugan (1984, 1985) studied the oxidative stability of olive oil stored in glass and poly(ethylene) (PE) plastic bottles and concluded that glass provides better protection from oxidation than plastic bottles. Kaya et al. (1993) studied the effect of permeability and transparency of glass and polyethylene terephthalate (PET) bottles on the SL of sunflower and olive oil by monitoring the peroxide values. The authors concluded that colored glass was superior to clear glass and PET in terms of the protection provided for packaged olive oil. Tawfik and Huyghebaert (1999) also evaluated the effects of various plastic films [PET, polyvinyl chloride (PVC), polypropylene (PP), and polystyrene (PS)] on the stability of various vegetable oils compared to glass packaging. Besides the type of packaging material, different storage conditions (time and temperature) and the presence of antioxidants, either in the oils or in the plastics, were taken into account. The research highlighted that preservation in glass was better than in plastics after 20 days of storage, while after 2 months of storage the plastic bottles completely altered an oils characteristics. The ranking of oxidative stability in olive oil was glass > PVC = PET > PP > PS at both temperatures. Oxygen permeability, the level of natural antioxidants in oils, time, and storage temperature were found to be the main characteristics affecting oil stability. This study also proved that natural compounds directly incorporated into oils retarded the extent of oxidation, more than antioxidants in polymeric materials, even though a migration of the preservative agents from the packaging to the product was recorded.

Mass transport phenomena are of great importance to food packaging with plastics since a polymeric matrix is permeable to moisture, oxygen, carbon dioxide, nitrogen, and other low molecular weight compounds. Glass and metal packaging materials are not permeable to low molecular weight compounds. Hence, these last types of material do not allow the designer to optimize the barrier properties for various applications. Polymers can provide a wide range (by three or four orders of magnitude) of permeability for various applications, thus justifying studies aimed at ensuring adequate barrier protection (Masi and Paul 1982). For foods like olive oil that are sensitive to oxygen or moisture, barrier protection against low molecular weight compounds is the major function of packages in providing adequate protection (Brown 1992).

As was widely discussed in the overview presented by Kanavouras et al. (2006) on the packaging of olive oil, the use of mathematical modeling for the prediction of packaged olive oil SL is highly desirable. The validation of the simulations against known experimental results confirmed the value of the mathematical approach for a quick and accurate prediction of the SL of oxidation-sensitive products. However, a limited number of valuable mathematical models have been presented in the literature to predict the SL of packaged olive oil and to suggest new package designs after taking into consideration the role of oxygen, the geometrical and structural characteristics of plastic containers, and the volume of the oil (Coutelieris and Kanavouras 2006; Del Nobile et al. 2003a; 2003b; Dekker et al. 2002; Gambacorta et al. 2004; Kanavouras et al. 2004; Kanavouras and Coutelieris 2006). Del Nobile et al. (2003a) worked to design a plastic bottle to package olive oil using a mathematical approach that could predict the evolution of the profile of hydroperoxide and oxygen concentration in bottled virgin olive oil. The approach used by these authors was described in detail in Chap. 3 of the modeling section; see Chap. 3 for an exhaustive explication of the argument. Briefly, Del Nobile et al. (2003a) used two bottles containing oxygen scavengers: a plastic bottle in which the oxygen scavenger is uniformly dispersed through the bottle wall and a glass bottle internally coated with a polymer in which the oxygen scavenger is uniformly dispersed. The model was derived assuming the average hydroperoxide concentration as a measure of virgin olive oil quality (Satue et al. 1995), without considering the diffusion of the flavor compounds in the oil phase. The mathematical model was obtained by combining the mass balance equations of oxygen and hydroperoxides with that describing the rate of hydroperoxide formation and decomposition, by using a simplified version of the model proposed by Quast and Karel (1972) to describe lipid oxidation in potato chips. To derive the model, three main assumptions were made: (1) the bottle containing virgin olive oil can be represented by a cylinder composed by an outer shell (made of plastic or glass) and by an internal oil core, (2) oxygen mass diffusion occurs only in the radial direction and through the lateral surface, and (3) the diffusive mass flux of hydroperoxides through both the olive oil and the container wall were considered negligible. The developed model assessed the effect of oxygen diffusivity, the thickness of the plastic container, the presence of an oxygen scavenger in the container wall, and the concentration of oxygen in the oil prior to bottling on the quality decay kinetics of packaged olive oil. In particular, it was established that it is possible to obtain a quality decay kinetic as slow as that obtained for olive oil bottled in glass by increasing the barrier properties of the polymer used to manufacture the bottle. The same result cannot be obtained by increasing the thickness of the container wall due to the oxygen dissolved in the container wall. A quality decay kinetic slower than that found with glass bottles can be obtained by combining the oxygen scavenger with a biodegradable plastic blend based on starch-poly(caprolactone) or by bottling the oil in PET containers and reducing the oxygen concentration prior to bottling to 10 % of the equilibrium value, the oxygen's dissolution in oil prior to bottling the main factor causing quality decay during storage.

Even though the aforementioned model was used to demonstrate several advantageous aspects related to the design of plastic bottles for packaging virgin olive oil, it has some limitations. The limitations are a direct consequence of the use of empirical equations to describe hydroperoxide formation and breakdown reactions, and the consideration that oxygen diffusion takes place only in the radial direction (monodimensional model). Consequently, the proposed model cannot be used to predict the quality decay kinetics of small containers or to assess the influence of a bottle's geometrical factors on the quality decay kinetics of bottled oil. Therefore, to further assess the influence of some of the bottle's geometrical factors on the quality decay kinetics of virgin olive oil bottled in glass and plastic containers, a two-dimensional model was proposed (Del Nobile et al. 2003b). To that end, five geometrically different bottles were investigated. The first three differed in the volumetric capacity (i.e., 1, 1/2, and 1/4 L), while the latter two contained the same amount of oil (1 L) but differed in the capacity of bottle headspace. By simulating the behavior of bottled virgin olive oil, the mathematical model was able to predict the evolution of hydroperoxides and oxygen concentration profile inside bottled virgin olive oil during storage and to assess the influence of the bottle's shape and size on the quality decay kinetic of virgin olive oil in glass and plastic containers. The approach proposed by Del Nobile et al. (2003b) demonstrated that the extent to which the geometrical factors of packaging affect the quality loss of oil depends on the material used to make the bottle and on the initial value of the oxygen partial pressure in the bottle headspace. To control the oxidation kinetics during the storage of bottled oil, it may be useful to use well-designed plastic bottles and innovative plastic materials containing an oxygen scavenger. Moreover, it could be highly advisable to perform the bottling operations under a nitrogen atmosphere to reduce the oxygen pressure in the bottle headspace. By comparing the monodimensional with the two-dimensional model, the authors observed that the two mathematical approaches had a similar predictive ability in the case of the bottle with a volumetric capacity of 1 L; most probably this is due to the fact that during the first stage of storage the oxidation reaction rate depends primarily on the oxygen dissolved in the oil prior to bottling. The differences between the quality decay kinetics predicted by the two models was ascribed to the fact that the two-dimensional model takes into account the axial mass flux coming from the top and the bottom of the PET bottle, as well as the presence of the oxygen in the packaging headspace, whereas in the case of glass bottles, the difference between the predictions of the two models is less marked than that observed in the case of plastic bottle because for glass bottles there is no mass flux from the bottom of the bottle.

Coutelieris and Kanavouras (2006) proposed an advanced approach to describing the mass transport from and to the oil phase through various packaging materials under several temperatures and light conditions. Contrary to previous predictive models also reported in the literature, the authors incorporated the mass transport of the most oxidation-characteristic compounds due to diffusion, as well as the interactions of the packaging materials with flavor compounds. The model was developed by considering a set of mass transport equations describing the chemical reactions occurring in the oil phase, as well as the diffusion of oxygen in the oil phase and through the packaging material. In addition, the lack of probability that the packaged olive oil would reach the end of its SL during a certain time was also estimated and proposed as a quality reduction indicator. The essential data employed to validate the proposed model focused on the oxidative deterioration of extra virgin olive oil, packaged in 0.5 L glass and plastic containers (PET and PVC) and stored at 15, 30, and 40 °C under fluorescent light or dark conditions for 1 year. The evolution of hexanal over time was used as the main indicator of the oxidative alterations taking place inside the oil phase over time. Results highlighted that olive oil stored at the lowest temperature under light contained the lowest amounts of hexanal, while when stored in the dark using any packaging material it had comparable amounts of hexanal. The model satisfactorily fits the experimental data, except in oil stored in the dark due to the very low concentrations of hexanal, thus representing a valid instrument for SL modeling of packaged oil.

6.3   Packaging of Wine

Wine is one of the most ancient fermented beverages. It is an alcoholic product made from fermentation, mainly sustained by yeasts of *Saccharomices* spp. of must. The must is generated by the mechanical compression and filtration of *Vitis vinifera* berries. The extent of the fermentation process and, consequently, the ethanol content are defined on the basis of the amount of sugar in the grape used. The wine color depends mainly on the pigments of the berry skin and on the winery technology (maceration process), while the flavor is affected by the grape cultivar and the fermentation process. According to the European Committee (EEC Regulation 822/87), table wine is defined as "wine other than quality wine," excluding those wines with an appellation of origin. The term designates inexpensive, mild-flavored wines, with a SL generally limited to 6 months. Wines are not perishable products, but their organoleptic properties can be seriously affected by oxygen and light (Escudero et al. 2002). When wine is not well protected through sufficient sulfitation, the presence of high oxygen content could promote an acetic bacteria attack and the subsequent development of acetic acid and ethyl acetate. However, chemical oxidation can also produce significant sensory modifications in wine flavor and color, mainly consisting of a loss of aromatic freshness and the appearance of brown precipitates of condensed phenolic material (Singleton 1987; Cheynier et al. 1989; Zurbano et al. 1995; Simpson 1982). The capacity of wine to take up oxygen and to withstand oxidation is roughly measured by the total content of phenols, phenols being the major substrates for oxidation in wine. Obviously, phenol content decreases as a consequence of oxidation. Finally, wine oxidation can also involve the appearance of aldehydes, such as (E)-2-hexenal (Culleré et al. 2007), methional, and phenylacetaldehyde (Ferreira et al. 2003), which affect product quality.

Glass containers sealed using a cork are usually preferred for bottling all types of wine because glass is a material with a high barrier to gases and vapors and is stable in time, transparent, and highly recyclable. However, glass's use as long-lasting package for short-lived products and its high cost make it less than ideal, which has prompted an increasing demand for inexpensive and alternative solutions for wine bottling. Synthetic plastic corks with an elastic property and low oxygen permeability are one of the most innovative developments of wine packaging (Lopez et al. 2007).

To date, the substitution of glass with plastic materials for wine packaging has been proposed. Pasquarelli (1983) evaluated the permeability to water, moisture, and oxygen of PET, PVC, PS, and PP of various grades, suggesting that some grade of these polymers or polymer-coated cardboard could replace aluminum-based multilayer structures for wine packaging. Due to its transparency, excellent mechanical properties, low price, low weight, and good oxygen barrier properties, PET is the most investigated plastic material for food applications (Del Nobile et al. 2003a; Gambacorta et al. 2004; Ros-Chumillas et al. 2007; Zygoura et al. 2004). Moreover, PET serves as a relatively good barrier against permeation of flavor compounds due to the biaxial orientation of its molecules (Van Lune et al. 1997). Ough (1987) reported a study on the use of PET containers for wine, showing that 4 L PET and 3 L PET with an extra high barrier Saran layer were as effective in wine preservation as glass up to 10 months. No significant change in the content of some volatile esters was detected, although the esters did tend to be slightly lower in samples stored in PET containers. Boidron and Bar (1988) revealed that the quality decay of PVC-packaged wines was higher than that packaged in PET bottles. Buiatti et al. (1996, 1997) compared red and white wine preservation in PET, wine boxes, multilayer cartons, and glass by monitoring the amount of oxygen dissolved, the total phenols, the value of absorbance at 420 and 520 nm, the sulfur dioxide, and the volatile acidity, showing that multilayer cartons were particularly efficient at preserving wine. Recently, due to increasing interest in enhancing packaging functionality (Brody et al. 2001; Smith et al. 1995; Verimeiren et al. 1999), active systems were also applied with success to wine. Giovanelli and Brenna (2007) experimented with the use of PET with an oxygen scavenger and found that active plastic bottles could be very effective against oxidative processes in young wines and could replace glass containers for the entire storage period. Mentana et al. (2009) also studied the quality decay of Apulia table wines as affected by PET with a 4 % oxygen scavenger package (OxSc-PET, 1 L monolayer PET with thickness 0.3 mm). Oxygen transfer through the plastic bottles into the wine, flavor sorption on the package, and migration from the packaging generally represent the three main factors affecting wine quality. Obviously, when food-grade packaging is used, only the gas barrier properties and scalping phenomenon must be taken into account to assess product acceptability (Giovanelli and Brenna 2007; Macías et al. 2001; Sajilata et al. 2007). For this reason, in the work of Mentana et al. (2009), classic enological parameters, including the anthocyanin fraction and the volatile fraction, together with sensory properties were monitored over 7 months on red and white local table wines, stored at 15–18 °C in the dark, to assess the chemical changes related to oxidation and flavor scalping. The results showed that PET with oxygen scavenger demonstrated a behavior much closer to that of glass bottles in preserving wines than PET without scavenger. In particular, red wine bottled in PET showed significant losses in most of the volatile compounds, including alcohols, acids, and esters. OxSc-PET containers showed scalping effects on a smaller number of products and to a lesser extent, probably due to the presence of the oxygen scavengers that affect packaging material properties, both decreasing its polarity and preventing oxygen permeation. In white wine bottled in both PET and OxSc-PET plastic packages, a smaller number of compounds were involved in scalping. From a sensory point of view, all wines were acceptable up to 7 months, although the ranks assigned to each wine revealed statistically significant differences between PET and OxSc-PET containers in favor of the active packaging.

The increasing consumer demand over the last decade for low-environmental-impact polymeric materials (An et al. 2001; Petersen et al. 1999) has justified the numerous efforts to investigate applications of transparent, cheap, and effective biodegradable or recyclable materials to foodstuffs (Conte et al. 2009; Del Nobile et al. 2009; Haugaard et al. 2002, 2003; Frederiksen et al. 2003; Holm and Mortensen 2004). In this context, biodegradable packaging has also been proposed for wine applications. In particular, polylactic acid (PLA), a material currently obtained from renewable resources and considered a safe form of food packaging with regard to the migration of harmful components, has been used (Pati et al. 2010; Conn et al. 1995; Ljungberg et al. 2002). In the work of Pati et al. (2010), the authors evaluated the quality of carbonic maceration wines packaged in PLA bottles compared to PET and glass. Carbonic maceration is a preyeast fermentation treatment of intact products: grapes stored in a closed container consume O_2 and produce CO_2 by respiration. The anaerobic treatment, lasting for up to 2 weeks or longer, causes intracellular fermentation before starting conventional yeast fermentation. Carbonic maceration yields light red wines with low tannins, intense color, and fresh, fruity flavors, which are not suitable for aging and should be consumed early (Jackson 2000). As previously reported, the SL of a wine is directly related to the oxygen content to which it is exposed (Escudero et al. 2002) and to the sensory modifications to flavor, color, and appearance (Benítez et al. 2006; Gómez et al. 1995). Therefore, classic enological parameters, including volatile fraction, and sensory attributes were monitored by the aforementioned authors over a 4-month storage period. Total acidity, pH alcoholic grade, and total phenol values were shown to be unaffected by packaging; no statistically significant difference between wine packaged in polymeric containers and wine packaged in glass was observed. As expected, total SO_2 content decreased during storage in all packages but was significantly higher in the control and in the PET-bottled wine than in the PLA, most probably due to the higher permeability to oxygen of PLA, which provoked a more pronounced SO_2 depletion. Volatile acidity showed values in PLA-stored samples that were significantly higher compared to PET and glass. Oxidation phenomena were also shown to be more relevant in PLA-bottled wine than in wine stored in bottles of PET and glass, as suggested by the changes recorded in terms of volatile fraction. From a sensory point of view, glass, PET, and PLA were preferred in this order (95 % confidence level). Figure 6.1 shows a comparison of scores obtained at 3 and 4 months of storage in terms of appearance, taste, and odor

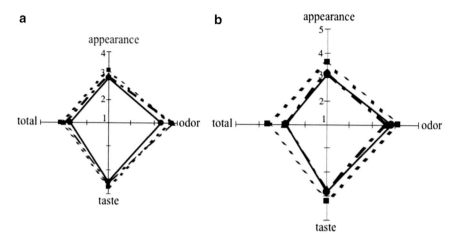

Fig. 6.1 Spider diagram relevant to sensory analysis performed with 35 untrained judges. Scores of appearance, odor, taste, and total acceptability were given ratings from 1 to 5: 1 *very good*, 2 *good*, 3 *acceptable*, 4 *poor*, 5 *very poor*. Samples were considered unacceptable when the mean score was above 3. Glass (●), PET (▲), and PLA (■) packaged wines for (**a**) 3-month storage and (**b**) 4-month storage

of the differently packaged carbonic maceration wine. The PLA container, although causing a faster wine quality loss compared to PET and glass, was judged to be suitable as a replacement for glass or traditional plastic for up to 3-months in storage, and PLA has the additional advantage of being an eco-friendly material.

The action of the Productive Activities Ministry on March 24, 2005, published in the ministry's official journal on April 5, which authorizes the bottling of waters in small 125, 250, or 500 mL containers, has opened a window not only on the use of a new typology of packaging material but also on the volume of the container. Life outside the home, lunches in cafes, and business trips have forced people to think about possible uses for monodose containers in cafes and on airplanes. To make small plastic materials suitable and acceptable in wine packaging, instead of heavy and fragile glass containers, further study is necessary (Boidron and Bar 1988). The information gained from earlier works dealing with the packaging of wine in plastic bottles was not entirely applicable to monodose applications due to changes in the surface-to-volume ratio, as observed with respect to oil storage by Del Nobile et al. (2003b). For this reason the Department of Food Science at the University of Foggia, in cooperation with PET Engineering, a company in Treviso that manufactures PET packaging, carried out a research project on the sensory impact on a young wine preserved for 4 months in glass, PET, and PET with extra barrier properties in 200 mL monodose bottles (Mentana et al. 2008). This research demonstrated the need to broaden the application of economically viable and well-accepted plastic materials to wine and to replace heavy and fragile glass containers with different materials. The project started outwith a study conducted by PET Engineering on the feasability of the container. After monitoring the sensory properties of packaged novello wine, the research group assessed the plastic bottles had an efficay similar to that of glass. The authors summed up the results stating that the success of wine in PET packaging depends on consumer demand and on the specific needs of a given wine. A single solution for all types of wine is inconceivable, but cooperation with customers is key to finding the best solution for each specific need.

References

An JH, Kim HS, Chung DJ, Lee DS (2001) Thermal behaviour of poly(ε-caprolactone)-poly (ethylene glycol)-poly(ε-caprolactone) tri-block copolymers. J Mater Sci 36:715–722

Angelo AJS (1996) Lipid oxidation in foods. Crit Rev Food Sci Nutr 36:175–224

Askar A (1999) Flavor changes during processing and storage of fruit juices-II. Interaction with packaging materials. Fruit process 11:432–439

Baldioli M, Servili M, Perretti G, Montedoro GF (1996) Antioxidant activity of tocopherols and phenolic compounds of virgin olive oil. J Am Oil Chem Soc 73:1589–1593

Benítez P, Castro R, Natera R, Barroso CG (2006) Changes in the polyphenolic and volatile content of "Fino" sherry wine exposed to high temperature and ultraviolet and visible radiation. Eur Food Res Technol 222:302–309

Boidron JN, Bar M (1988) Effects of bottle material on wine in small-volume containers. Connaissance de la Vigne et du Vin 22:73–83

Brody AL, Strupinsky ER, Kline LR (2001) Active packaging for food applications. Technomic Publishing Company, Pennsylvania

Brown WE (1992) Plastics in food packaging – properties, design, and fabrication. Marcel Dekker, New York

Buiatti S, Celotti E, Zironi R, Ferrarini R (1996) Wine storability in containers other than glass. Industrie delle Bevande 25:409–413

Buiatti S, Celotti E, Ferrarini R, Zironi R (1997) Wine packaging for market in containers other than glass. J Agric Food Chem 45:2081–2084

Cheynier V, Basire N, Rigaud J (1989) Mechanism of trans-caffeoyltartaric acid and catechin oxidation in model solution containing grape polyphenoloxidase. J Agric Food Chem 4:1069–1071

Conn RE, Kolstad JJ, Borzelleca JF, Dixler DS, Filer LJ Jr, LaDu BN Jr (1995) Safety assessment of polylactide (PLA) for use as a food-contact polymer. Food Chem Toxicol 33:273–283

Conte A, Gammariello D, Di Giulio S, Attanasio M, Del Nobile MA (2009) Active coating and modified atmosphere packaging to extend the shelf life of Fior di latte cheese. J Dairy Sci 92:887–894

Coutelieris FA, Kanavouras A (2006) Experimental and theoretical investigation of packaged olive oil: development of a quality indicator based on mathematical predictions. J Food Eng 73:85–92

Culleré L, Cacho J, Ferreira V (2007) An assessment of the role played by some oxidation-related aldehydes in wine aroma. J Agric Food Chem 55:876–881

Dalpasso L (1991) Experience with packaging of edible oils, wine and vinegar in PVC bottles. Ras Im Con 12:15–16

Dekker M, Kramer M, Van Beest M, Luning P (2002) In: Proceedings of 13th IAPRI conference on packaging, East Lansing, 1, pp 297–303

Del Nobile MA, Mensitieri G, Nicolais L, Masi P (1997) The influence of the thermal history on the shelf life of carbonated beverages bottled in plastic containers. J Food Eng 34:1–13

Del Nobile MA, Ambrosino ML, Sacchi R, Masi P (2003a) Design of plastic bottles for packaging of virgin olive oil. J Food Sci 68:170–175

Del Nobile MA, Bove S, La Notte E, Sacchi R (2003b) Influence of packaging geometry and material properties on the oxidation kinetic of bottles virgin olive oil. J Food Eng 57:189–197

Del Nobile MA, Conte A, Scrocco C, Brescia I (2009) New strategies for minimally processed cactus pear packaging. Inn Food Sci Emerg Technol 10:356–362

Escudero A, Asensio E, Cacho J, Ferreira V (2002) Sensory and chemical changes of young white wines stored under oxygen. An assessment of the role played by aldehydes and some other important odorants. Food Chem 77:325–331

Fayoux SC, Seuvre AM, Voilley AJ (1997) Aroma transfer in and through plastic packagings: orange juice and d-limonene. A review. Part I: orange juice aroma sorption. Packag Technol Sci 10:69–82

Ferreira ACS, Hogg T, De Pinho PG (2003) Identification of key odorants related to the typical aroma of oxidation-spoiled white wines. J Agric Food Chem 51:1377–1381

Frankel EN (1998) Lipid oxidation. The Oily Press, Dundee, p 303

Franz R, Huber M, Piringer O-G, Damant AP, Jickells SM, Castle L (1996) Study of functional barrier properties of multilayer recycled poly(ethylene terephthalate) bottles for soft drinks. J Agric Food Chem 44:892–897

Frederiksen CS, Haugaard VK, Poll L, Miquel Becker E (2003) Light-induced quality changes in plain yoghurt packaged in polylactate and polystyrene. Eur Food Res Technol 217:61–69

Gambacorta G, Del Nobile MA, Tamagnone P, Leonardi M, Faccia M, La Notte E (2004) Shelf-life of extra virgin olive oil stored in packages with different oxygen barrier properties. Ital J Food Sci 16:417–428

Giovanelli G, Brenna OV (2007) Oxidative stability of red wine stored in packages with different oxygen permeability. Eur Food Res Technol 226:169–179

Gómez E, Martínez A, Laencina J (1995) Prevention of oxidative browning during wine storage. Food Res Int 28:213–217

Haugaard VK, Weber CJ, Danielsen B, Bertelsen G (2002) Quality changes in orange juice packaged in materials based on polylactate. Eur Food Res Technol 214:423–428

Haugaard VK, Danielsen B, Bertelsen G (2003) Impact of polylactate and poly(hydroxybutyrate) on food quality. Eur Food Res Technol 216:233–240

Holm VK, Mortensen G (2004) Food packaging performance of polylactate (PLA). In: Conference proceedings, 14th IAPRI world conference on packaging, Stockholm, 13–16 June 2004

IOOC (International Olive Oil Council) (1984) International trade standards applying to olive oil and olive residue olis. COI/T. ISNC No 1

Kanavouras A, Coutelieris FA (2006) Shelf-life predictions for packaged olive oil based on simulations. Food Chem 96:48–55

Kanavouras A, Hernandez-Munoz P, Coutelieris F, Selke S (2004) Oxidation derived flavor compounds as quality indicators for packaged olive oil. J Am Oil Chem Soc 81:251–257

Kanavouras A, Hernandez-Munoz P, Coutelieris F (2006) Packaging of olive oil: quality issues and shelf life predictions. Food Rev Int 22:381–404

Kaya A, Tekin AR, Oner MD (1993) Oxidative stability of sunflower and olive oil: comparison between a modified active oxygen method and long term storage. Lebensm Wiss Technol 26:464–468

Keys A (1970) Coronary heart disease in seven countries. Circulation 44(suppl 1)

Keys A, Menotti A, Karvonen JM, Aravanis C, Blackburn H, Buzzina R, Djorjevic BS, Dontas A, Fidanza F, Keys HN, Kromhout D, Nedeljkovic S, Punsar S, Seccareccia F, Toshima H (1986) The diet and 15-year death rate in the seven countries study. Am J Epidemiol 124:903

Kiritsakis AK, Dugan LR (1984) Effect of selected storage conditions and packaging materials on olive oil quality. J Am Oil Chem Soc 61:1868–1870

Kiritsakis AK, Dugan LR (1985) Studies in photooxidation of olive oil. J Am Oil Chem Soc 62:892–896

Kiritsakis AK, Kanavouras A, Kiritsakis K (2002) Chemical analysis, quality control and packaging issues of olive oil. Eur J Lipid Sci Technol 104:628–638

Kuchel L, Brody AL, Wicker L (2006) Oxygen and its reactions in beer. Packag Technol Sci 19:25–32

Jackson RS (2000) In Wine science – principles, practice, perception. Academic Press. San Diego

Labuza TP (1971) Kinetics of lipid oxidation in foods. CRC Crit Rev Food Technol 10:355–405

Licciardello F, Del Nobile MA, Spagna G, Muratore G (2009) Scalping of ethyloctanoate and linalool from a model wine into plastic films. LWT- Food Sci Technol 42:1065–1069

Ljungberg N, Andersson T, Wesslén B (2002) Film extrusion and film weldability of poly(lactic acid) plasticized with triacetine and tributyl citrate. J Appl Polym Sci 88:3239–3247

Lopez P, Saucier C, Teissedre PL, Glories Y (2007) Main routes of oxygen ingress through different closures into wine bottles. J Agric Food Chem 55:5167–5170

Macías VMP, Pina IC, Rodríguez LP (2001) Factors influencing the oxidation phenomena of sherry wine. Am J Enol Viticult 52:151–155

Masi P, Paul DR (1982) Modelling gas transport in packaging applications. J Membr Sci 22:137–151

Mastrobattista G (1990) Effect of light on extra virgin olive oils in different types of glass bottles. Ital J Food Sci 3:191–195

Mentana A, Pati S, Gambacorta G, Del Nobile MA, La Notte E (2008) La bottiglia monodose è di plastica. Food packag 19:24–28

Mentana A, Pati S, La Notte E, Del Nobile MA (2009) Chemical changes in Apulia table wines as affected by plastic packages. LWT- Food Sci Technol 42:1360–1366

Nielsen T, Jagerstad M (1994) Flavour scalping by food packaging. Trends Food Sci Technol 5:353–356

Ough CS (1987) Use of PET bottles for wine. Am J Enol Vitic 38:100–104

Pasquarelli O (1983) Plastic container for table wines. Plastic 14(10):61–64

Pati S, Mentana A, La Notte E, Del Nobile MA (2010) Biodegradable poly-lactic acid package for the storage of carbonic maceration wine. LWT- Food Sci Technol 43:1573–1579

Petersen K, Nielsen PV, Bertelsen G, Lawther M, Olsen MB, Nilsson NH, Mortensen G (1999) Potential of bio-based materials for food packaging. Trends Food Sci Technol 10:52–68

Quast DG, Karel M (1972) Computer simulation of storage life of foods undergoing spoilage by two interacting mechanisms. J Food Sci 37:679–683

Riquet AM, Wolff N, Laoubi S, Vergnaud JM, Feigenbaum A (1998) Food and packaging interactions: determination of the kinetic parameters of olive oil diffusion in polypropylene using concentration profile. Food Addit Contam 15:690–700

Robertson G-L (1993) Food packaging. Marcel Dekker, New York, pp 338–380

Ros-Chumillas M, Belissario Y, Iguaz A, López A (2007) Quality and shelf life of orange juice aseptically packaged in PET bottles. J Food Eng 79:234–242

Sajilata MG, Savitha K, Singhal RS, Kanetkar VR (2007) Scalping of flavours in packaged foods. Compr Rev Food Sci Food Saf 6:17–35

Satue MT, Huang SW, Frankel EN (1995) Effect of natural antioxidants in virgin olive oil on oxidative stability refines, bleached and deodorized olive oil. J Am Oil Chem Soc 72:1131–1137

Simpson RF (1982) Factors affecting oxidative browning of white wines. Vitis 21:233–239

Singleton VL (1987) Oxygen with phenols and related reactions 347 in musts, wines, and model systems: observations and practical implications. Am J Enol Vitic 38:69–77

Smith JP, Hoshino J, Abe Y (1995) Interactive packaging involving sachet technology. Active food packaging. London: Blackie Academic and Professional 143–173

Tawfik MS, Huyghebaert A (1999) Interaction of packaging materials and vegetable oils: oil stability. Food Chem 64:451–459

Tsimidou M, Papadopoulos G, Boskou D (1992) Phenolic compounds and stability of virgin olive oil. Part I. Food Chem 45:141–144

Van Lune FS, Nijssen LM, Linssen JPH (1997) Absorption of methanol and toluene by polyester-based bottles. Packag Technol Sci 10:221–227

Verímeiren L, Devlieghere F, Van Beest M, De Kruijf N, Debevere J (1999) Developments in the active packaging of foods – laser-induced surface modification of polymers. Trends Food Sci Technol 10:77–86

Viola P (1970) In facts in the human diet. International Olive Oil Council, Madrid

Zurbano FP, Ferreira V, Peña C, Escudero A, Serrano F, Cacho JF (1995) Prediction of oxidative browning in white wines of their chemical composition. J Agric Food Chem 43:2813–2817

Zygoura P, Moyssiadi T, Badeka A, Kondyli E, Savvaidis I, Kontominas MG (2004) Shelf life of whole pasteurized milk in Greece: effect of packaging material. Food Chem 87:1–9

Chapter 7
Minimally Processed Food: Packaging for Quality Preservation

7.1 Introduction

Minimally processed crops consist of washed, cut, and packaged fruits and vegetables. Consumer preferences are increasingly geared toward ready-to-use fruits and vegetables, thus implicating a great effort of research in this area. Food marketplace evolves new products and changing trends, and fresh-cut products remain at the top of the list of products meeting the needs of busy consumers. The value of fresh-cut products lies in the primary characteristics of freshness and convenience. However, operations such as peeling, cutting, shredding, and slicing greatly increase tissue damage of fresh-cut produce (Soliva-Fortuny and Martìn-Belloso 2003; Martin-Diana et al. 2007; Olivas et al. 2007) and may result in several biochemical deteriorations such as browning, off-flavor production, and loss of texture and degraded nutritional value and microbial quality of products (Giménez et al. 2003; Watada and Qi 1999). Among the various alterations in transpiration, enzymatic activity, water loss, flavor, aroma, volatile profiles, and microbial proliferation, for most products the accelerated respiration rate represents the main factor that provokes food unacceptability (Ragaert et al. 2007; Rico et al. 2007). Thus, strategies aimed at slowing down respiration activity are generally successful for such food (Fonseca et al. 2002). Usually, a relevant aspect to be taken into account for maintaining the quality of horticultural commodities is the choice of appropriate packaging system (Del Nobile et al. 2008a; 2009a). Various head-space conditions can be achieved in a package depending on the interactions between the respiratory activity of the packaged produce and gas transfer through the polymeric film. The superposition of both processes leads to an increase in CO_2 and a reduction in O_2 in the package headspace. In fact, Smith et al. (1987) reported that, as a result of produce respiration, matching of commodity characteristics to film permeability could result in the passive evolution of an appropriate atmosphere within a sealed package. Therefore, the choice of film mass barrier properties is a

M.A. Del Nobile and A. Conte, *Packaging for Food Preservation*,
Food Engineering Series, DOI 10.1007/978-1-4614-7684-9_7,
© Springer Science+Business Media New York 2013

key factor in obtaining optimum modification of the atmosphere and avoiding extremely low levels of O_2 or high levels of CO_2, which could induce anaerobic metabolism with the possibility of off-flavor generation or the risk of anaerobic microorganism proliferation (Beaudry 2000; Watkins 2000). Various predictive models of the respiration rate of minimally processed food have been developed to optimize packaging characteristics or optimal gas composition (Rocculi et al. 2006; Del Nobile et al. 2007). A modified atmosphere in a package (MAP) can be created either passively by the product (passive MAP) or intentionally, introducing gas mixtures into the package (active MAP). In passive MAP, the respiring product is placed in a polymeric package and sealed hermetically. Only the respiration of the product and the gas permeability of the film influence the change in gaseous composition of the environment surrounding the product. If the product respiration characteristics are properly matched to the film permeability values, a beneficial modified atmosphere can be passively created within the package. Conversely, in the case of active MAP, the atmosphere surrounding the product is removed or replaced before the package is sealed. The gas headspace may change during storage in MAP, but there is no additional manipulation of the internal environment (Mastromatteo et al. 2010). The selection of a proper gas combination that avoids the usual transient state before reaching the equilibrium state of the gas in the bag can reduce the respiration rate during the transient state, thereby promoting product preservation and, consequently, shelf life prolongation (Lee et al. 1996; Rodriguez-Aguilera and Oliveira 2009). MAP can be interpreted as a dynamic system with two gas fluxes, the respiration rate of the fresh product and the gas exchange through the barrier (Van de Velde and Kiekens 2002). In general, gas compositions inside a MAP package are low in O_2 and high in CO_2, depending primarily on temperature, product weight, respiration rate, O_2 and CO_2 transmission rates, and total respiring surface area (Mahajan et al. 2007). The natural variability of raw material and its dynamic response to processing and storage conditions may render it impossible to identify a truly optimal atmosphere using general empirical methods, suggesting that significant advances in the packaging of minimally processed food may require the development of complex mathematical models that incorporate the dynamic response of products to the environment (Jacxsens et al. 1999; Saltveit 2003). MAP and low-temperature storage are usually not sufficient to extend the shelf life of cut produce because the excessive physiological stress and increased susceptibility to microbial spoilage caused by processing operations reduce significantly the shelf life (Mastromatteo et al. 2010). Under natural conditions, the outer layer of plant tissue consists of a hydrophobic surface, providing a natural barrier to microorganisms (Lund 1992). Due to damage to the surface, nutrients are released from the plant tissue and can be used by microorganisms. It is well established that fresh produce may contain a high contamination level after harvest. It may range between 3 and 7 log units depending on the season and type of produce. Pathogenic microorganisms associated with fresh vegetables can cause severe foodborne disease outbreaks. For this reason, pretreatment with antimicrobial or antioxidant compounds is highly advised to further contribute to product preservation. In recent years there has been considerable pressure by consumers to reduce or eliminate chemically synthesized additives in foods. Thus, efforts have been made to find natural alternatives to the currently used additives to prevent bacterial and fungal growth in minimally processed fruits and vegetables. Antimicrobial compounds can be incorporated into the packaging material, coated on the surface of the packaging film, or added in a sachet into the package to be released during storage (active packaging). Another possibility for natural carrier compounds is to incorporate the compound into an edible coating, either by dipping or spraying the food or by adding the active compounds directly in the food-making process (Mastromatteo et al. 2009, 2010; Siracusa et al. 2008). Natural antimicrobials can be defined as molecules of natural origin that are nontoxic for humans, environmentally safe, inexpensive, and widely available (Burt 2004). The main sources of these compounds are plants (e.g., plant secondary metabolites in essential oils and phytoalexins), microorganisms (e.g., bacteriocins and organic acids) and animals (e.g., lysozyme from eggs, chitosan from crab and shrimp shell wastes, and transferrins from milk) (Bari et al. 2005; Meyer et al. 2002).

The variability of raw materials and the dynamic response to processing and storage conditions may render it impossible to deal with all crop types in a single chapter. Some case studies were taken into account as examples of three important minimally processed categories: (1) fruit with a hydrophobic skin resistant to microbial attack (grape), (2) a vegetable very susceptible to the browning process (lampascioni), and (3) a vegetable with a high susceptibility to yellowing (broccoli). The approaches, which are based on scientific principles, were presented and widely discussed in each case study.

7.2 Table Grape

Table grape is a nonclimacteric fruit with severe problems during postharvest. Gray mold, caused by *Botrytis cinerea*, is the principal cause of decay of table grapes both in the field and after harvest. Additionally, the deterioration during storage is also attributed to water loss, stem drying, browning, and softening of berries (Valero et al. 2006). Grape quality depends on numerous factors, such as harvest period, climatic and soil conditions, cultural practices, degree of ripening, variety, and sanitary conditions (Conte et al. 2012; Palacios et al. 1997; Keller et al. 1998). Like many other nonclimacteric fruits, grape has a relatively low rate of physiological activity that generally increases with the respiration rate. In fact, the pattern of respiration can be divided into two typologies: climacteric and nonclimacteric. This latter pattern of respiration belongs to some fruits like grape and to most vegetables and maintains a relatively constant level throughout storage, whereas for climacteric fruit, respiration increases to a maximum with storage time (Lee et al. 1996).

Fungicides, and in particular SO_2, are the primary means to control postharvest diseases, but due to the side effects of synthetic chemicals on human health, their use has been limited (Zoffoli et al. 1999; LaTorre et al. 2002; Lydakis and Aked 2003). Numerous effective, safe, and environmentally sound alternative strategies have been developed and applied to various table grape cultivars for reducing quality losses. Hot water immersion treatments (Fallik 2004) and dips in ethanol (Lichter et al. 2002, 2003; Karabulut et al. 2004; Pesis 2005) or in chlorinated water (Ahvenainen 1996; Soliva-Fortuny and Belloso 2003) are the most widespread approaches in the literature. Del Nobile et al. (2008b) compared all three of these treatments on minimally processed table grape (*Vitis vinifera* cv. Italia). In particular, the effects of hot water at 55 °C for 5 min and dip in ethanol (50 % solution) for 5 min and in chlorinated water (20 ml L^{-1}) for 15 min were tested on the quality decay of packaged fruit. Due to increased emphasis on environmentally friendly films and the need for commercial breakthroughs in green polymers, the grape clusters were packaged in biodegradable bags made up of polyesters (F1, thickness 18 μm, and F2, thickness 25 μm). The quality decay of fresh table grape stored at 5 °C was assessed by monitoring the respiration rate, the cell load of the main spoilage microorganisms (total mesophilic viable count, acid lactic bacteria, yeasts, and molds), and the product appearance for a period of approximately 30 days. The respiration rate was calculated using a mathematical approach proposed by the same authors also for other minimally processed food (Del Nobile et al. 2006, 2007), as described in detail in Part II of this book. Microbiological quality of fruit was monitored by plate count, and visual quality was evaluated by image analysis (Image Pro Plus v. 6, Media Cybernetics, Silver Spring, MD, USA) of grape clusters at different storage times. The results of principal component analysis revealed differences between samples in terms of respiration rate. The grape treated in hot water showed a higher respiration rate than the other two treated samples, regardless of packaging adopted. In fact, the total amount of oxygen consumed by the packaged produce treated in hot water was higher than that of the other samples. Considering that oxygen depletion is directly related to the extent of metabolic activities (senescence level) associated with product respiration (Böttcher et al. 2003), it was possible to underline the differences between the three treatments applied to clusters prior to packaging. As regards the effects on microbial quality, all strategies appeared effective in controlling spoilage, even though a very slow contamination was recorded on the grapes during storage. This experimental evidence could be ascribed to the very hard and smooth skin of berries, which protects the inner tissue from yeasts and fungal invasion and prevents the discharge of the inner juice, thus limiting the availability of nutrients for the natural microflora (Thournas and Katsoudas 2005). Trivial differences between color parameters of grape samples were also recorded. As an example of images analyzed to assess the visual quality, Fig. 7.1 reports a picture of various grape clusters at 1 day of storage. Results underlined a lack of specific trend at each sampling time, thus suggesting that differences could be attributed to the natural evolution of raw material (Nunan et al. 1998); the process conditions adopted in the work cannot be considered responsible for retaining visual quality. To sum up, both packagings exerted the same protection, while ethanol seemed to be a slightly better solution for minimally processed table grape compared to the other two treatments.

The use of natural compounds of plant origin also appears to be a very promising means to control the postharvest decay of grape. Among the natural compounds,

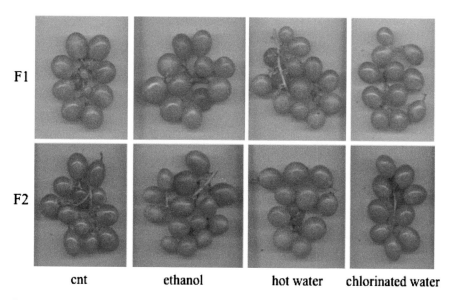

Fig. 7.1 Images of clusters of untreated and treated table grape in ethanol, hot water, and chlorinated water, packaged in two different bio-based films (F1, thickness 18 μm and F2, thickness 25 μm), after 1 day of storage. Cnt = control sample without any treatment; ethanol = samples treated with ethanol; hot water = samples treated with hot water; chlorinated water = samples treated with chlorinated water

plant essential oils (EOs) are attracting interest for their potential as natural food preservatives (Lanciotti et al. 2004; Tripathi and Dubey 2004). They are generally recognized as safe (GRAS), and many of them display a wide spectrum of antimicrobial activity, with a capacity to control foodborne pathogens and spoilage bacteria associated with ready-to-eat vegetables (Mastromatteo et al. 2009). EOs also show antioxidant activity due to a number of mechanisms, such as free-radical scavenging, hydrogen donation, singlet oxygen quenching, metal-ion chelation, and acting as substrates for radicals (Mastromatteo et al. 2010). Various attempts to apply active compounds (eugenol, thymol, menthol, and eucalyptol) to cultivars of table grapes have been made (Guillèn et al. 2007; Valverde et al. 2005; Valero et al. 2006). The EOs were placed on a sterile gauze inside bags, preventing contact with fruit, and the bags were rapidly sealed to minimize vaporization. Fruit quality parameters highlighted that samples treated with eugenol, thymol, or menthol showed benefits in terms of reduced weight loss, delayed color changes, and maintenance of fruit firmness compared to the control. Stems remained green in treated samples but became brown in the control. Conversely, samples packaged with eucalyptol behaved worse than untreated grape, generating off-flavors, loss of quality, and stem browning.

Among the nontoxic replacements of synthetic fungicides, hypobaric treatments (Romanazzi et al. 2001), storage with high CO_2 (Retamales et al. 2003), and biocontrol agents (Wilson and Wisniewski 1989; Schena et al. 2002) have also

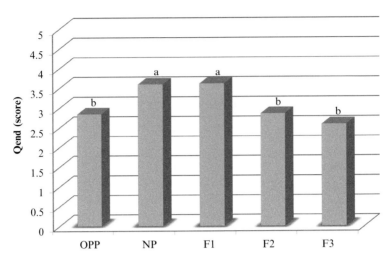

Fig. 7.2 Values of general visual quality of minimally processed table grape at end of observation period (35 days) (Qend). The values were calculated by fitting Eq. 7.1 to the sensory data. Statistically significant differences ($p < 0.05$) between samples are indicated by different letters. OPP = grape sample packaged in oriented polypropylene, thickness 20 μm; NP = grape sample packaged in a multilayer film obtained by laminating a nylon layer with a polyolefin-based film, thickness 95 μm; F1 = grape sample packaged in a biodegradable film, thickness 100 μm; F2 = grape sample packaged in a biodegradable film, thickness 50 μm; F3 = grape sample packaged in a biodegradable film, thickness 35 μm

attracted considerable attention. However, due to a lack of consistency in the application of these last-named methodologies as standalone treatments and due to injuries to grape barriers, their use is very limited (Valero et al. 2006).

The effects of headspace packaging conditions (passive MAP) for retaining the quality of grape are well known in the literature (Artés-Hernandez et al. 2006; Kou et al. 2007; Del Nobile et al. 2008b, 2009a). In various scientific researches, table grapes were packaged in several plastic nonperforated and perforated films, especially based on polyethylene and polypropylene, giving rise to a wide range of modified headspace gas concentrations, depending on the respiration rate of the grape cultivar and on the mass transport properties of the selected polymeric film (Martìnez-Romero et al. 2003). Films with lower gas permeance generate a lower headspace oxygen concentration and a higher headspace carbon dioxide concentration, which in turn allow reducing the respiratory activity of the product and, consequently, the senescence level, with a positive effect on the visual quality (Del Nobile et al. 2009a). For example, Fig. 7.2 shows the values of the visual quality recorded on table grape packaged in five different films, three biodegradable films (F1, thickness 100 μm; F2, thickness 50 μm; F3, thickness 35 μm), an oriented polypropylene film (OPP, thickness 20 μm), and a multilayer film obtained by laminating a nylon layer with a polyolefin-based film (NP, thickness 95 μm) (Del Nobile et al. 2009a). To quantitatively determine the influence of film permeability on grape sensory decay, a first-order-type equation was fitted to the preceding data:

$$Q(t) = \frac{Q_{end} - Q_{in} \cdot \exp(-k \cdot t_{end})}{1 - \exp(-k \cdot t_{end})}$$
$$+ \left[Q_{in} - \frac{Q_{end} - Q_{in} \cdot \exp(-k \cdot t_{end})}{1 - \exp(-k \cdot t_{end})} \right] \cdot \exp(-k \cdot t), \qquad (7.1)$$

where $Q(t)$ is the grape general visual quality at time t, Q_{end} is the packaged grape general visual quality at the end of the observation period, Q_{in} is the initial value of the packaged grape general visual quality, k is the kinetic constant, t_{end} is the time at the end of the experimental observation (35 days), and t is the storage time. As can be inferred from Fig. 7.2, Q_{end} values are all close to 3 (the acceptability limit), thus suggesting that all tested films successfully preserved grapes for at least 35 days. However, a slight higher score was observed for the two films with the highest gas barrier properties, thus allowing the authors to conclude that the headspace reached in carbon dioxide played a major role in preserving the packaged grape quality (Crisosto et al. 2002; Retamales et al. 2003).

Sometimes, the protective effects of the packaging materials were further enhanced by the combination with antimicrobial or antioxidant compounds directly incorporated into the packaging system (Lurie et al. 2006; Moyls et al. 1996; Valero et al. 2006; Valverde et al. 2005; Zoffoli et al. 1999).

Although the beneficial effects of the active MAP technique for retaining the quality of fruits and vegetables are well known (Rodriguez-Aguilera and Oliveira 2009), studies dealing with the packaging of grapes under active MAP conditions are lacking. To the best of our knowledge, the sole experimental work on active MAP of table grape is the study of Costa et al. (2011). The authors compared the effects of passive and three active modified headspace conditions (5:3:92, 10:3:87, and 15:3:82 O_2:CO_2:N_2) on fruit packaged in three types of bags, based on OPP, with different thicknesses (20, 40, and 80 μm, respectively). During a prolonged storage period at refrigerated temperature (5 °C), the headspace gas concentrations, mass loss, microbiological stability, and sensory acceptability were assessed. By means of a proper data elaboration, the SL of the unpackaged minimally processed produce and of grape packaged under the tested conditions was calculated. Results are displayed in Fig. 7.3, where it is possible to observe that all packaging films significantly prevented grape decay compared to unpackaged product. The best results were recorded in the thickest OPP sealed in air, which assured a SL of more than 2 months. It is worth noting that the active MAPs were not found to be effective for SL prolongation due to the fast equilibrium of gas reached in the bags and due to a more pronounced product dehydration. Active MAP, theoretically, offers the possibility of improving the product quality and freshness and increasing the SL. However, as stated earlier, fruits and vegetables are unlike other foods because they consume oxygen and produce carbon dioxide when packaged, giving rise to a modification of the headspace gas composition. In both cases the gas headspace may change during storage, but there is no additional manipulation of the internal environment. The selection of an initial gas combination different from air might only succeed at avoiding the transient state, which is usually generated before

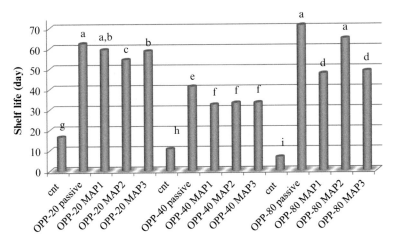

Fig. 7.3 Values of shelf life (day) of packaged and unpackaged minimally processed table grape. Statistically significant differences ($p < 0.05$) between samples are indicated by different letters. Cnt = unpackaged sample; OPP-20 passive = grape sample packaged in oriented polypropylene, thickness 20 μm; OPP-20 MAP1 = grape sample packaged in OPP 20 μm under 5:3:92 O_2:CO_2: N_2 MAP conditions; OPP-20 MAP2 = grape sample packaged in OPP 20 μm under 10:3:87 O_2: CO_2:N_2 MAP conditions; OPP-20 MAP3 = grape sample packaged in OPP 20 μm under 15:3:82 O_2:CO_2:N_2 MAP conditions; OPP-40 passive = grape sample packaged in oriented polypropylene, thickness 40 μm; OPP-40 MAP1 = grape sample packaged in OPP 40 μm under 5:3:92 O_2: CO_2:N_2 MAP conditions; OPP-40 MAP2 = grape sample packaged in OPP 20 μm under 10:3:87 O_2:CO_2:N_2 MAP conditions; OPP-40 MAP3 = grape sample packaged in OPP 40 μm under 15:3:82 O_2:CO_2:N_2 MAP conditions; OPP-80 passive = grape sample packaged in oriented polypropylene, thickness 80 μm; OPP-80 MAP1 = grape sample packaged in OPP 80 μm under 5:3:92 O_2:CO_2:N_2 MAP conditions; OPP-80 MAP2 = grape sample packaged in OPP 80 μm under 10:3:87 O_2:CO_2:N_2 MAP conditions; OPP-80 MAP3 = grape sample packaged in OPP 80 μm under 15:3:82 O_2:CO_2:N_2 MAP conditions

gas equilibrium in the bag is reached. By reducing the transient state it is possible to reduce the respiration rate, thereby promoting product preservation and, consequently, SL prolongation (Lee et al. 1996). Due to the variability of raw materials, it is impossible to generalize the effects of active MAP on fresh-cut produce; a dedicated study is necessary for each specific product (Saltveit 2003).

7.3 Lampascioni (*Muscari comusum*)

Lampascioni (*Muscari comosum*) is a typical product of the Apulia region in southern Italy. The edible portion is a walnut-sized bulb of a wild hyacinth, not an onion, with a characteristic bitter taste (Chiej 1984). Figure 7.4 shows an example of lampascioni after peeling. The cooked bulbs, preserved in oil, are also used as a relish. Lampascioni is a highly perishable vegetable that deteriorates rapidly after peeling due to the browning process and weight loss. The rate of

Fig. 7.4 Example of peeled lampascioni

browning determines its sensory SL (Jacxsens et al. 2002, 2003); therefore, the understanding of the processes leading to product changes is essential in developing better approaches to minimizing detriment and, hence, improving quality and SL (Toivonen and Brummell 2008). Enzymatic browning is a particular problem in fruit with a white flesh. It is generally assumed to be a direct consequence of polyphenol oxidase (PPO) and peroxidase (POD) action on polyphenols to form quinones, which ultimately polymerize to produce the browning appearance of fresh-cut produce. Browning is associated with the loss of membrane integrity, which occurs during tissue deterioration and senescence. Thus, membrane stability is potentially a major factor controlling browning of fruit and vegetable slices. Based on the working mechanisms, browning inhibitors can be categorized into six groups comprising reducing agents, acidulants, chelating agents, complexing agents, enzyme treatments, and enzyme inhibitors (McEvily et al. 1992; Lamikanra and Watson 2001; Rojas-Graü et al. 2008; Saltveit 2000). Among all inhibitors tested, reducing agents (e.g., ascorbic acid and its derivatives, cysteine and glutathione, citric acid alone or in combination, as a substitute for sulphites) have been proven effective in controlling browning (Albanese et al. 2007; Cocci et al. 2006; Gimenez et al. 2003; Lee et al. 2003; Rocculi et al. 2004). It has been suggested that ascorbic acid provides protection by acting as an oxygen scavenger, thus inhibiting PPO reactions, and by self-oxidation to prevent oxidation by phenol compounds (Mayer and Harel 1979; Rico et al. 2007). In contrast, citric acid may have a dual inhibitory effect on PPO by reducing the pH and by chelating the copper to the enzyme-active site (McEvily et al. 1992). A calcium-ascorbate-based formula has been widely used by the fresh-cut-apple industry (Mastromatteo et al. 2009). The incorporation of antioxidant agents such as N-acetylcysteine and glutathione into alginate- and gellan-based coatings has helped to prevent fresh-cut apples, papayas, and pears from browning (Oms-Oliu ET AL. 2008; Rojas-Graü et al. 2007). Alginate-based edible coating formulated with citric acid (1 %) was used to prolong the SL of fresh-cut Madrigal artichokes packed in a polyester-based biodegradable film (Del Nobile et al. 2009b). Among the renewable sources available to produce edible coatings, the polysaccharide-based materials are the most widespread because they are abundant, cheap, and easy to use (Devlieghere et al. 2004; Lee et al. 2003).

Conte et al. (2009) implemented a technique for keeping lampascioni ready to cook based on a combination of a treatment applied to the product prior to packaging and the selection of a proper polymeric film. To accomplish the aim of the work, two different treatments were tested: dipping in a solution containing citric acid (1 %) and calcium chloride (8 %) and coating with sodium alginate (5 %) in combination with citric acid (1 %) and calcium chloride (8 %). Treated and untreated samples were packaged in two types of polymeric film: a commercially available OPP film and a polyester-based biodegradable film. The environmentally friendly films may represent suitable materials for ready-to-use products due to their low barrier properties. However, their use is still very limited due to the high conversion cost and small amounts available (Del Nobile et al. 2006, 2008a; Muratore et al. 2006). Packaged lampascioni samples were stored at 5 °C for approximately 20 days. During refrigerated storage, respiration rate, microbial populations, pH, weight loss, and visual quality were monitored. Results showed that coated lampascioni packaged in the biodegradable film were preserved best, thus confirming the suitability of the active coating, combined with the gas barrier properties of film, in controlling the detrimental phenomena involved in product deterioration. In particular, it was found that both dip and coating reduced the respiration activity of lampascioni, even if the coating treatment was more effective. In fact, it was also found in the literature that edible coatings based on whey protein concentrate, carrageenan, or polysaccharide/lipid formulation, in combination with antibrowning agents, reduced the respiration rate of sliced apples (Lee et al. 2003; Wong et al. 1994). Conte et al. (2009) also determined that coating treatment, more than packaging, affected microbial proliferation and reduced water loss, as reported in the literature for other vegetables (Park 1999; Ragaert et al. 2007). To assess product visual quality, the experimental data, recorded by a sensory analysis carried out with a trained panel of judges, were fitted with the following Gompertz equation, as reparameterized by Corbo et al. (2006). The same approach was also described in Chap. 8 for dairy products (mozzarella cheese):

$$VQ(t) = VQ_{min} - A \cdot \exp\left\{-\exp\left\{\left[(\mu_{max} \cdot 2.71) \cdot \frac{\lambda - VQAL}{A}\right] + 1\right\}\right\}$$
$$+ A \cdot \exp\left\{-\exp\left\{\left[(\mu_{max} \cdot 2.71) \cdot \frac{\lambda - t}{A}\right] + 1\right\}\right\}, \tag{7.2}$$

where $VQ(t)$ is the lampascioni visual quality at time t, VQ_{min} is the minimum value of $VQ(t)$ to consider the packaged lampascioni still acceptable from a general appearance point of view, A is related to the difference between the initial $VQ(t)$ value and that attained at equilibrium, μ_{max} is the maximal decay rate, λ is the lag time, $VQAL$ is the lampascioni visual quality acceptability limit [defined as the storage time at which $VQ(t)$ reaches its acceptability limit (VQ_{min})], and t is the storage time. Results from the fitting procedure are displayed in Fig. 7.5. The data

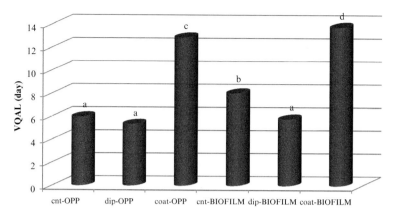

Fig. 7.5 Mean values of visual quality acceptability limit (VQAL) of lampascioni, obtained by fitting Eq. 7.2 to the experimental data of the sensory analysis. VQAL is defined as the time (day) at which lampascioni visual quality reaches its acceptability limit (score = 3). Statistically significant differences ($p < 0.05$) between samples are indicated by different letters. Cnt-OPP = control sample packaged in OPP; Dip-OPP = dipped sample packaged in OPP; Coat-OPP = coated sample packaged in OPP; Cnt-BIOFILM = control sample packaged in bio-based film; Dip- BIOFILM = dipped sample packaged in bio-based film; Coat- BIOFILM = coated sample packaged in bio-based film

underline the fact that no substantial differences were found between the *VQAL* values of uncoated samples, whereas the coating treatment significantly improved visual acceptability in both types of packaging, with a slight better effect in the biodegradable polymeric matrix. The reason for these findings could be the lower headspace oxygen concentration recorded in the eco-friendly bags that, combined with the action of citric acid and the protection of the coating, further contributed to slow down the browning process, thereby prolonging product acceptability.

7.4 Broccoli

Broccoli is composed of a number of immature floral buds (florets) and thick, fleshy flower branchlets or stalks attached to the central plant axis, which is collectively named the head. For consumption, heads are harvested while they are totally immature. Whole broccoli is available to consumers wrapped in 10–14 μm thickness PVC films and generally have a SL of 3–4 weeks in air at 0 °C (Makhlouf et al. 1989) but only a few days at 20 °C (Wang 1977). The yellowing of florets is the most evident sign of broccoli deterioration, resulting from the degradation of chlorophyll (Ceponis et al. 1987). As expected, fresh-cut broccoli has a shorter SL than whole broccoli; this is due to the high respiration rate of the vegetable. The quality loss is mainly caused by surface dehydration and loss of green color with yellowing and opening of florets, loss of stem firmness, development of undesirable odors, and soft rots (Shewfelt et al. 1984; Berrang et al. 1990; Forney et al. 1993). In response to the plant tissue alterations, microorganisms proliferate, further contributing to product SL reduction (Nguyen and Carlin 1994; Brackett 1989).

Broccoli has been intensively studied by various researchers who used several techniques to extend its postharvest life; however, most works deal with intact broccoli. Refrigeration (Gillies and Toivonen 1995), modified atmosphere (Lipton and Harris 1974; Makhlouf et al. 1989; Rai et al. 2008), heat (Costa et al. 2005), ethanol (Suzuki et al. 2004), UV-C radiation (Costa et al. 2006), and the use of 1-methylcyclopropene (1-MCP) (Able et al. 2002) are examples of preservation methods applied to the whole product. On the other hand, modified atmosphere (Rakotonirainy et al. 2001), UV-C (Lemoine et al. 2007), and ethanol vapor (Han et al. 2006) were used to delay the postharvest senescence of fresh-cut broccoli.

The scientific literature dealing with this topic shows that broccoli can benefit from 1 % to 2 % O_2 and 5 % to 10 % CO_2 at low temperatures. A low O_2 level (0.5–2 %) or an excess of CO_2 concentration, combined with temperature fluctuations, may result in off-odors (volatile compounds containing sulfur) that make the product unmarketable (Forney and Rij 1991; Makhlouf et al. 1989). Consequently, the choice of film mass transport properties is a key factor in obtaining the optimal atmospheric modification and in avoiding extremely low levels of O_2 or high levels of CO_2 (Conte et al. 2008, 2009; Del Nobile et al. 2009a). Depending on the interactions between the respiratory activity of the packaged produce and gas transfer through the polymeric matrix, different headspace conditions can be achieved in the package, thus exerting different effects on the product. For this reason, Lucera et al. (2011) investigated the suitability of packaging films with different gas barrier properties on the quality decay of fresh-cut broccoli (*Brassica oleracea* L., var. Italica). The authors first made screened several polymeric matrices to choose the most appropriate one for packaging minimally processed broccoli; then the effects of the selected films on the main quality indices of the fresh-cut produce were investigated. In particular, amounts of about 100 g of broccoli florets were first packaged in films composed of OPP with three different thicknesses (20, 40, and 80 μm) and of PP (thickness 30 μm) with different microperforations (50, 20, 12, 9, and 7 μ holes per package, with diameter 70 μm). During refrigerated storage conditions, changes in the headspace gas composition were monitored in all the filled packages to evaluate the ability of the nonperforated and microperforated packaging films to create the optimum headspace conditions surrounding the fresh-cut crop. The gas permeability of all the films was calculated following a procedure similar to that proposed by Larsen et al. (2000, 2002). For more details on modeling concerns, the reader is invited to refer to the publication of Lucera et al. (2011). The recorded values of permeability were found to span two orders of magnitude, thus covering a wide range of mass transport properties. The results obtained from the headspace monitoring agreed with the gas transport properties of films. In fact, the OPP-based bags were found to serve as a high barrier for fresh-cut broccoli because they provoked

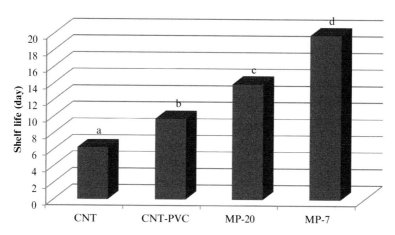

Fig. 7.6 Mean values of shelf life (day) recorded on fresh-cut broccoli packaged and unpackaged. The shelf life values were obtained by matching the microbiological and sensorial acceptability limits (MAL and SAL). Statistically significant differences ($p < 0.05$) between samples are indicated by different letters. Cnt = unpackaged control sample; Cnt-PVC = sample wrapped in PVC; MP-20 = sample packaged in microperforated package with 20 µ holes; MP-7 = sample packaged in microperforated package with 7 µ holes

a complete depletion of oxygen. In contrast, in the bag with the highest number of holes, gas concentrations remained fairly constant and florets appeared less green, browner, and noncompact. Therefore, among the various packaging systems, two microperforated films (with 7 and 20 holes per package) were chosen for the second trial. This last step was intended as a fine optimization, aimed at assessing the influence of the packaging film mass transport properties on the SL. To this end, approximately 100 g of broccoli florets were packaged in the two selected microperforated films. In addition, unpackaged fresh-cut florets of broccoli (CNT) and intact broccoli wrapped in PVC (thickness 12 µm) (CNT-PVC) were also stored in the same conditions as the controls. Headspace gas concentration, mass loss, and microbiological and sensory quality were monitored. The unpackaged product was affected by a significant mass loss (40 % in 10 days) compared to all other packaged samples. From a microbiological point of view, no substantial differences were recorded between the samples, whereas considerable differences were found on the sensory acceptability, odor and color being the main quality attributes for broccoli acceptability. Microbiological and sensory data were both fitted with a modified version of the Gompertz equation to calculate the microbial and sensory acceptability limits (MAL and SAL, respectively), according to the approach proposed in Chap. 8 for dairy fresh food. Results highlighted that all samples recorded a MAL higher than the storage time because during the entire observation period the threshold for microbial acceptance was never reached. Conversely, different SAL values were obtained. Since SL depends on both microbial and sensory quality, SAL coincided with product SL as the microbial load was always found below the threshold (Fig. 7.6). According to findings recorded on mass loss, the firmness was the main factor influencing the acceptability of unpackaged broccoli. A pronounced tissue yellowing was manifested on samples wrapped with PVC, thus compromising product acceptability after 10 days of storage. In contrast, fresh-cut broccoli packaged in microperforated films, especially film with seven holes per package, retained the green color characteristic of freshly harvested broccoli, with no symptoms of off-flavors at the moment of bag opening. The outcomes recorded by Lucera et al. (2011) suggest that the selection of the proper packaging is of crucial importance in creating headspace conditions able to guarantee the maintenance of sensory characteristics and delaying degradation processes.

References

Able AJ, Wong LS, Prasad A, O'Hare TJ (2002) 1-MCP is more effective on a floral brassica (Brassica oleracea var. italica l.) than a leafy brassica (Brassica rapa var. chinensis). Postharvest Biol Technol 26:147–155

Ahvenainen R (1996) New approaches in improving the shelf life of minimally processed fruit and vegetables. Trends Food Sci Technol 7:179–187

Albanese D, Cinquanta L, Di Matteo M (2007) Effects o fan innovative dipping treatment on the cold storage of minimally processed Annurca apples. Food Chem 105:1054–1060

Artés-Hernández F, Tomás-Barberán FA, Artés F (2006) Modified atmosphere packaging preserves quality of SO$_2$-free 'Superior seedless' table grapes. Postharvest Biol Technol 39:146–154

Bari ML, Ukuku DO, Kawasaki T, Inatsu Y, Isshiki K, Kawamoto S (2005) Combined efficacy of nisin and pediocin with sodium lactate, citric acid, phytic acid, and potassium sorbate and EDTA in reducing the *Listeria monocytogenes* population of inoculated fresh-cut produce. J Food Prot 68:1381–1387

Beaudry RM (2000) Responses of horticultural commodities to low oxygen: limits to the expanded use of modified atmosphere packaging. Hortic Technol 10:491–500

Berrang ME, Brackett RE, Beuchat LR (1990) Microbial, colour and textural qualities of fresh asparagus, broccoli, and cauliflower stored under controlled atmosphere. J Food Prot 53:391–395

Bottcher H, Gunther I, Kabelitz L (2003) Physiological postharvest responses of common Saint-John's worth herbs (Hypericum perforatum). Postharvest Biol Technol 29:342–350

Brackett RE (1989) Changes in the microflora of packaged fresh broccoli. J Food Qual 12:169–181

Burt S (2004) Essential oils: their antibacterial properties and potential applications in foods – a review. Int J Food Microbiol 94:223–253

Ceponis MJ, Cappellini RA, Lightner GW (1987) Disorder in cabbage, bunched broccoli and cauliflower shipment to the New York market, 1972–1985. Plant Dis 71:1151–1154

Chiej R (1984) Encyclopaedia of medicinal plants. MacDonald, London, pp 10–15

Cocci E, Rocculi P, Romani S, Dalla Rosa M (2006) Changes in nutritional properties of minimally processed apples during storage. Postharvest Biol Technol 39:265–271

Conte A, Conversa G, Scrocco C, Brescia I, Laverse J, Elia A, Del Nobile MA (2008) Influence of growing periods on the quality of baby spinach leaves at harvest and during storage as minimally processed produce. Postharvest Biol Technol 50:190–196

Conte A, Scrocco C, Brescia I, Del Nobile MA (2009) Packaging strategies to prolong the shelf life of minimally processed lampascioni (*Muscari comosum*). J Food Eng 90:199–206

Conte A, Mastromatteo M, Antonacci D, Del Nobile MA (2012) Influence of cultural practices and packaging materials on table grape quality. J Food Process Eng 35:701–707

Corbo MR, Del Nobile MA, Sinigaglia M (2006) A novel approach for calculating shelf life of minimally processed vegetables. Int J Food Microbiol 106:69–73

Costa ML, Civello PM, Chaves AR, Martìnez GA (2005) Effect of hot air treatments on senescence and quality parameters of harvested broccoli (Brassica oleracea L. var italica) heads. J Sci Food Agric 85:1154–1160

Costa L, Vicente AR, Civello PM, Chaves AR, Martìnez GA (2006) UV-C treatment delays postharvest senescence in broccoli florets. Postharvest Biol Technol 39:204–210

Costa C, Lucera A, Conte A, Mastromatteo M, Speranza B, Antonacci A, Del Nobile MA (2011) Effects of passive and active modified atmosphere packaging conditions on ready-to-eat table grape. J Food Eng 102:115–121

Crisosto CH, Garner D, Crisosto G (2002) Carbon dioxide-enriched atmosphere during cold storage limit losses from Botrytis but accelerate rachis browning of "Red globe" table grape. Postharvest Biol Technol 26:181–189

Del Nobile MA, Baiano A, Benedetto A, Massignan L (2006) Respiration rate of minimally processed lettuce as affected by packaging. J Food Eng 74:60–69

Del Nobile MA, Licciardello F, Scrocco C, Muratore G, Zappa M (2007) Design of plastic packages for minimally processed fruits. J Food Eng 79:217–224

Del Nobile MA, Conte A, Cannarsi M, Sinigaglia M (2008a) Use of biodegradable films for prolonging the shelf life of minimally processed lettuce. J Food Eng 85:317–325

Del Nobile MA, Sinigaglia M, Conte A, Speranza B, Scrocco C, Brescia I, Bevilacqua A, Laverse J, La Notte E, Antonacci D (2008b) Influence of postharvest treatments and film permeability on quality decay kinetics of minimally processed grapes. Postharvest Biol Technol 47:389–396

Del Nobile MA, Conte A, Scrocco C, Brescia I, Speranza B, Sinigaglia M, Perniola R, Antonacci D (2009a) A study on quality loss of minimally processed grapes as affected by film packaging. Postharvest Biol Technol 51:21–26

Del Nobile MA, Conte A, Scrocco C, Laverse J, Brescia I, Conversa G, Elia A (2009b) New packaging strategies to preserve fresh-cut artichoke quality during refrigerated storage. Innov Food Sci Emerg Technol 10:128–133

Devlieghere F, Vermeulen A, Debevere J (2004) Chitosan: antimicrobial activity, interactions with food components and applicability as a coating on fruit and vegetables. Food Microbiol 21:703–714

Fallik E (2004) Pre-storage hot water treatments (immersion, rinsing and brushing). Review. Postharvest Biol Technol 32:125–134

Fonseca SC, Oliveira FAR, Brecht JK (2002) Modeling respiration rate of fresh fruits and vegetables for modified atmosphere packages: a review. J Food Eng 52:99–119

Forney CF, Rij RE (1991) Temperature of broccoli florets at time of packaging influences package atmosphere and quality. Hortic Sci 24:1301–1303

Forney CF, Hildebrand PD, Saltveit ME (1993) Production of methanethiol by anaerobic broccoli and microorganisms. Acta Hortic 343:100–104

Gillies SL, Toivonen PMA (1995) Cooling method influences the postharvest quality of broccoli. Hortic Sci 30:313–315

Gimenez M, Olarte C, Sanz S, Lomas C, Echàvarri JF, Ayala F (2003) Relation between spoilage microbiological quality in minimally processed artichoke packaged with different films. Food Microbiol 20:231–242

Guillèn F, Zapata PJ, Martìnez-Romero D, Castillo S, Serrano M, Valero D (2007) Improvement of the overall quality of table grapes stored under modified atmosphere packaging in combination with natural antimicrobial compounds. J Food Sci 72:185–190

Han J, Tao W, Hao H, Zhang B, Jiang W, Niu T, Li Q, Cai T (2006) Physiology and quality responses of fresh-cut broccoli florets pretreated with ethanol vapor. J Food Sci 75:385–389

Jacxsens L, Devlieghere F, Debevere J (1999) Validation of a systematic approach to design equilibrium modified atmosphere packages for fresh-cut produce. Lebensm Wiss Technol 32:425–432

Jacxsens L, Devlieghere F, Debevere J (2002) Predictive modelling for packaging design: equilibrium modified atmosphere packages of fresh cut vegetables subjected to a simulated distribution chain. Int J Food Microbiol 73:331–341

Jacxsens L, Devlieghere F, Ragaert P, Vanneste E, Debevere J (2003) Relation between microbiological quality, metabolite production and sensory quality of equilibrium modified atmosphere packaged fresh-cut produce. Int J Food Microbiol 83:263–280

Karabulut OA, Mlikota Gabler F, Mansour M, Smilanick JL (2004) Postharvest ethanol and hot water treatments of table grapes to control gray mold. Postharvest Biol Technol 34:169–177

Keller M, Arnink KJ, Harazdina G (1998) Interaction of nitrogen availability during bloom and light intensity during veraison. I. Effects on grapevine growth, fruit development, and ripening. Am J Enol Vitic 49:333–340

Kou L, Luo Y, Wu D, Liu X (2007) Effects of mild heat treatment on microbial growth and product quality of packaged fresh-cut table grapes. J Food Sci 72:567–573

Lamikanra O, Watson MA (2001) Effects of ascorbic acid on peroxidase and polyphenoloxidase activities in fresh-cut cantaloupe melon. J Food Sci 66:1283–1286

Lanciotti R, Gianotti A, Patrignani F, Belletti N, Guerzoni ME, Gardini F (2004) Use of natural aroma compounds to improve shelf-life and safety of minimally processed fruits. A review. Trends Food Sci Technol 15:201–208

Larsen H, Kohler A, Magnus EM (2000) Ambient oxygen ingress rate method-an alternative method to Ox-Tran for measuring oxygen transmission rate of whole packages. Packag Technol Sci 13:233–241

Larsen H, Kohler A, Magnus EM (2002) Predicting changes in oxygen concentration in the headspace of nitrogen flushed packages by the ambient oxygen ingress rate method. Packag Technol Sci 15:139–146

LaTorre BA, Spadaro I, Rioja ME (2002) Occurrence of resistant strains of Botrytis cinerea to anilinopyrimidine fungicides in table grapes in Chile. Crop Prot 21:957–961

Lee L, Arult J, Lencki R, Castaigne F (1996) A review on modified atmosphere packaging and preservation of fresh fruit and vegetables: physiological basis and practical aspects – part II. Packag Technol Sci 9:1–17

Lee JY, Park HJ, Lee CY, Choi WY (2003) Extending shelf life of minimally processed apples with edible coatings and anti-browning agents. Lebensm Wiss Technol 36:323–329

Lemoine ML, Civello PM, Martìnez GA, Chaves AR (2007) Influence of postharvest UV-C treatment on refrigerated storage of minimally processed broccoli (Brassica oleracea var. Italica). J Sci Food Agric 87:1132–1139

Lichter A, Zuthhy Y, Sonego L, Dvir O, Kaplunov T, Sarig P, Ben-Arie R (2002) Ethanol controls post harvest decay of table grapes. Postharvest Biol Technol 24:301–308

Lichter A, Zhou H-W, Vaknin M, Dvir O, Zutchi Y, Kaplunov T, Lurie S (2003) Survival and responses of Botrytis cinerea after exposure to ethanol and heat. J Phytopathol 151:553–563

Lipton WJ, Harris CM (1974) Controlled atmosphere effects on the market quality of stored broccoli. J Am Soc Hortic Sci 99:200–205

Lucera A, Costa C, Mastromatteo M, Conte A, Del Nobile MA (2011) Fresh-cut broccoli florets shelf life as affected by packaging film mass transport properties. J Food Eng 102:122–129

Lund BM (1992) Ecosystems in vegetable foods. J Appl Bacteriol 73:115–126

Lurie S, Pesis E, Gadiyeva O, Feygenberg O, Ben-Arie R, Kaplunov T, Zatahy Y, Litcher A (2006) Modified ethanol atmosphere to control decay of table grapes during storage. Postharvest Biol Technol 42:222–227

Lydakis D, Aked J (2003) Vapour heat treatment of Sultanina table grapes. I: control of Botrytis cinerea. Postharvest Biol Technol 27:109–116

Mahajan PV, Oliveira FAR, Montanez JC, Frias J (2007) Development of user-friendly software for design of modified atmosphere packaging for fresh and fresh-cut produce. Innov Food Sci Emerg Technol 8:84–92

Makhlouf J, Castaigne F, Arul J, Willemot C, Gosselin A (1989) Long-term storage of broccoli under controlled atmosphere. Hortic Sci 24:637–639

Martin-Diana AB, Rico D, Frias JM, Barat JM, Henehan GTM, Barry-Ryan C (2007) Calcium for extending the shelf life of fresh whole and minimally processed fruits and vegetables: a review. Trends Food Sci Technol 18:210–218

Martìnez-Romero D, Guìllen F, Castillo S, Valero D, Serrano M (2003) Modified atmosphere packaging maintains quality of table grapes. J Food Sci 68:1838–1843

Mastromatteo M, Conte A, Del Nobile MA (2009) Preservation of fresh-cut produce using natural compounds. Review article. Stewart Postharvest Rev 4:4

Mastromatteo M, Conte A, Del Nobile MA (2010) Combined use of modified atmosphere packaging and natural compounds for food preservation. Review article. Food Eng Rev 2:28–38

Mayer MA, Harel E (1979) Review: polyphenoloxidases in plants. Phytochemistry 18:193

McEvily AJ, Iyengar R, Otwell WS (1992) Inhibition of enzymatic browning in foods and beverages. Crit Rev Food Sci Nutr 32:253–273

Meyer AS, Suhr KI, Nielsen P (2002) Natural food preservatives. In: Ohlsson T, Bengtsson N (eds) Minimal processing technologies in the food industries. Woodhead Publishing Ltd, Cambridge, pp 80–100

Moyls AL, Sholberg PL, Gaunce AP (1996) Modified atmosphere packaging of grapes and strawberries fumigated with acetic acid. Postharvest Biol Technol 31:414–416

Muratore G, Lanza CM, Baiano A, Tamagnone P, Nicolosi C, Del Nobile MA (2006) The influence of using different packaging on the quality decay kinetics of Cuccìa. J Food Eng 73:239–245

Nguyen-the C, Carlin F (1994) The microbiology of minimally processed fresh fruits and vegetables. Crit Rev Food Sci Nutr 34:371–401

Nunan KJ, Sims IM, Bacic A, Robinson SP, Fincher GB (1998) Changes in cell wall composition during ripening of grape berries. Plant Physiol 118:783–792

Olivas GI, Mattinson DS, Barbosa-Canovas GV (2007) Alginate coatings for preservation of minimally processed 'Gala' apples. Postharvest Biol Technol 45:89–96

Oms-Oliu G, Odriozola-Serrano I, Soliva-Fortuny R, Martìn-Belloso O (2008) Antioxidant content of fresh-cut pears stored in high-O_2 active package compared with conventional low-O_2 active and passive modified atmosphere packaging. J Agric Food Chem 56:932–940

Palacios VM, Nebot-Sanz E, Perèz-Rodrìguez L (1997) Use of factor analysis for the characterization and modelling of maturation of Palominograpes in the Jerez region. Am J Enol Vitic 48:317–322

Park HJ (1999) Development of advanced edible coatings for fruits. Trends Food Sci Technol 10:254–260

Pesis E (2005) The role of the anaerobic metabolites, acetaldehyde and ethanol, in fruit ripening, enhancement of fruit quality and fruit deterioration. Postharvest Biol Technol 37:1–19

Ragaert P, Devlieghere F, Debevere J (2007) Role of microbiological and physiological spoilage mechanisms during storage of minimally processed vegetables. Postharvest Biol Technol 44:185–194

Rai DR, Tyagi SK, Jha SN, Mohan S (2008) Qualitative changes in the broccoli (Brassica oleracea var. italica) under modified atmosphere packaging in perforated polymeric film. J Food Sci Technol 45:247–250

Rakotonirainy AM, Wang Q, Padua GW (2001) Evaluation of zein films as modified atmosphere packaging for fresh broccoli. J Food Sci 66:1108–1111

Retamales J, Defilippi BG, Arias M, Castillo P, Manrìquez D (2003) High-CO_2 controlled atmospheres reduce decay incidence in Thompson Seedless and Red Globe table grapes. Postharvest Biol Technol 29:177–182

Rico D, Martín-Diana AB, Barat JM, Barry-Ryan C (2007) Extending and measuring the quality of fresh-cut fruit and vegetables: a review. Trends Food Sci Technol 18:373–386

Rocculi P, Romani S, Dalla Rosa M (2004) Evaluation of physico-chemical parameters of minimally processed apples packed in non-conventional modified atmosphere. Food Res Int 37:329–335

Rocculi P, Del Nobile MA, Romani S, Baiano A, Dalla Rosa M (2006) Use of a simple mathematical model to evaluate dipping and MAP effects on aerobic respiration of minimally processed apples. J Food Eng 76:334–340

Rodríguez-Aguilera R, Oliveira JC (2009) Review of design engineering methods and applications of active and modified atmosphere systems. Food Eng Rev 1:66–83

Rojas-Graü MA, Tapia MS, Rodrìguez FJ, Carmona AJ, Martín-Belloso O (2007) Alginate and gellan based edible coatings as support of antibrowning agents applied on fresh-cut Fuji apple. Food Hydrocoll 21:118–127

Rojas-Graü MA, Tapia MS, Martìn-Belloso O (2008) Using polysaccharide-based edible coatings to maintain quality of fresh-cut Fuji apples. Lebensm Wiss Technol 41:139–147

Romanazzi G, Nigro F, Ippolito A, Salerno M (2001) Effect of short hypobaric treatments on postharvest rots of sweet cherries, strawberries and table grapes. Postharvest Biol Technol 22:1–6

Saltveit ME (2000) Wound induced changes in phenolic metabolism and tissue browning are altered by heat shock. Postharvest Biol Technol 21:61–69

Saltveit ME (2003) Is it possible to find an optimal controlled atmosphere? Postharvest Biol Technol 27:3–13

Schena L, Nigro F, Pentimone I, Ligorio A, Ippolito A (2002) Control of postharvest rots of sweet cherries and table grapes with endophytic isolates of Aureobasidium pullulans. Postharvest Biol Technol 30:209–220

Shewfelt RL, Heaton EK, Batal KM (1984) Non-destructive colour measurement of fresh broccoli. J Food Sci 49:1612–1613

Siracusa V, Rocculi P, Romani S, Dalla Rosa M (2008) Biodegradable polymers for food packaging: a review. Trends Food Sci Technol 19:634–643

Smith S, Geeson J, Stow J (1987) Production of modified atmosphere in deciduous fruits by the use of films and coatings. Hortic Sci 22:772–776

Soliva-Fortuny RC, Martìn-Belloso O (2003) New advances in extending the shelf-life of fresh-cut fruit: a review. Trends Food Sci Technol 14:341–353

Suzuki Y, Uji T, Terai H (2004) Inhibition of senescence in broccoli florets with ethanol vapor from alcohol powder. Postharvest Biol Technol 31:177–182

Thournas VH, Katsoudas E (2005) Mould and yeast flora in fresh berries, grapes and citrus fruit. Int J Food Microbiol 105:11–17

Toivonen PMA, Brummell DA (2008) Biochemical bases of appearance and texture changes in fresh-cut fruit and vegetables. Postharvest Biol Technol 48:1–14

Tripathi P, Dubey NK (2004) Exploitation of natural products as an alternative strategy to control post harvest fungal rotting of fruit and vegetables. A review. Postharvest Biol Technol 32:235–245

Valero D, Valverde JM, Martinez-Romero D, Guillen F, Castillo S, Serrano M (2006) The combination of modified atmosphere packaging with eugenol or thymol to maintain quality, safety and functional properties of table grapes. Postharvest Biol Technol 41:317–327

Valverde JM, Guillen F, Martinez-Romero D, Castillo S, Serrano M, Valero D (2005) Improvement of table grapes quality and safety by the combination of modified atmosphere packaging (MAP) and eugenol, menthol, or thymol. J Agric Food Chem 53:7458–7464

Van de Velde K, Kiekens P (2002) Biopolymers: overview of several properties and consequences on their applications. Polym Test 21:433–442

Wang CY (1977) Effect of aminoethoxy analog of rhizobitoxine and sodium benzoate on senescence of broccoli. Hortic Sci 12:54–56

Watada A, Qi L (1999) Quality of fresh-cut produce. Postharvest Biol Technol 15:201–205

Watkins CB (2000) Responses of horticultural commodities to high carbon dioxide as related to modified atmosphere packaging. Hortic Technol 10:501–506

Wilson CL, Winiewski ME (1989) Biological control of post-harvest diseases of fruits and vegetables. An emerging technology. Annu Rev Phytopathol 27:425–441

Wong WS, Tillin SJ, Hudson JS, Pavlath AE (1994) Gas exchange in cut apples with bi-layer coatings. J Agric Food Chem 42:2278–2285

Zoffoli JP, Latorre BA, Rodríguez EJ, Aldunce P (1999) Modified atmosphere packaging using chlorine gas generators to prevent *Botrytis cinerea* on table grapes. Postharvest Biol Technol 15:135–142

Chapter 8
Innovations in Fresh Dairy Product Packaging

8.1 Introduction

Cheese is obtained by coagulation of milk by bacterial fermentation, rennet, or acid. There are many cheese varieties, and they can be grouped on the basis of moisture content into three main categories: hard cheese (moisture content 20–42 %), semihard cheese (moisture content 44–55 %), and soft cheese (moisture content approximately 55 %). The shelf life (SL) of cheese varies from a few days to several months, depending on the variant. In particular, the SL of high-moisture fresh cheese is very short due to the evolution of mechanisms linked to chemical, biochemical, and physical processes. In particular, microorganisms (psychrotrophs, molds, and yeasts) and enzymes that are characteristic of food or derived from the surrounding environment set up various detrimental reactions. It is well known that as each mechanism advances, relevant modifications occur from sensory, nutritional, and safety points of view (Eliot et al. 1998). Evidence exists that the hygienic characteristics of milk have a direct effect on the quality of cheese (Albenzio et al. 2001, 2004). The SL of fresh milk-derived products is insufficient to reach high-income markets, and it is often incompatible with distribution criteria. However, these products have a potentially high earning power both because of their gastronomic importance and because they meet consumer demands. Indeed, high-income consumers greatly appreciate products closely linked to their area of origin that can evoke the history and culture of their production region. The improvement and spread of milk and dairy products provide income to many people and promote the social and economic growth of the producing areas.

Current technologies for preservation and SL extension of cheese include the heat treatment of milk and the use of proper packaging conditions. Pasteurization kills Gram-negative spoilage bacteria and inactivates lipoprotein lipase, which catalyzes the breakdown of milk triglycerides into free fatty acids with the development of off-odors. Provided the process operations after pasteurization run aseptically, only heat-resistant spore formers and thermostable enzymes can act

M.A. Del Nobile and A. Conte, *Packaging for Food Preservation*,
Food Engineering Series, DOI 10.1007/978-1-4614-7684-9_8,
© Springer Science+Business Media New York 2013

to restrict product SL. While pasteurization destroys most harmful bacteria initially present in raw milk, the processed product must also be protected from recontamination. Safety measures beyond pasteurization, such as exhaustive milk testing, handling, and transport protocols, further validate dairy product safety. Dairy processors rely on refrigerated storage, safe processing, and handling procedures to prevent recontamination and ensure product wholesomeness. An area of concentration is the search for scientific and technological solutions that might assist dairy processors in inhibiting the growth of microorganisms (Byrne and Bishop 1998). This includes new processing technologies, such as high-pressure processing, pulsed electric fields, irradiation, and carbon dioxide injection (O'Reilly et al. 2002), as well as the introduction of active compounds that have preservative properties, sometimes produced by the same microorganisms used in dairy cultures. As other processes such as irradiation and high-pressure technologies gain US Food and Drug Administration (FDA) approval, dairy processors will have an even greater array of preventative tools available to them. Other scientific advances, such as the addition of natural active compounds and the use of modified atmosphere packaging (MAP) (Subramaniam 1993) will help dairy processors ensure that the production lots being released to distribution for consumption are the safest dairy foods possible (MAP means that air in the package headspace is modified with another gas composition). The proper gas combinations required to prolong SL and retain the sensory quality depend on the type of cheese, and in particular on the moisture and fat content (Conte et al. 2009; Del Nobile et al. 2009b).

Due to the increasing consumer demand for food with health-promoting qualities, more efforts could be aimed at providing safe and high-quality products by maintaining intact the traditional production technology of dairy foods. Cheese makers rarely like to innovate their cheese-making procedures because generally they use traditional protocols of milk processing. Therefore, research attention focuses on the potential efficacy of new packaging systems to fulfill actual consumer requirements. To this end, packaging containing natural extracts with antimicrobial properties represents the latest innovation. Recently, natural antimicrobial agents have attracted the attention of researchers because they represent an advantage in terms of food safety and a valid alternative to synthetic food additives used to inhibit the principal detrimental phenomena. Several natural substances, such as lysozyme, chitosan, and natural extracts from plants, appear to be effective compounds in controlling microbial proliferation in dairy applications (Altieri et al. 2005; Sinigaglia et al. 2008; Gammariello et al. 2008). As reported in other chapters of this section, the combination of active compounds with MAP further enhances their effectiveness (Conte et al. 2009), even though the success of packaging in the specific case of dairy products is dependent on several parameters, such as the type of cheese, the initial microbial contamination, the process, and the storage conditions.

The future of the cheese marketplace will be enhanced as more of the basic chemical reactions are correlated to specific aspects of cheese making and thoroughly embraced by all levels of quality control and product development. Research in dairy food safety ranges from innovative processing technologies to rapid testing of finished products. Many of these safety controls focus on preventing and controlling microbial contamination during manufacturing processes. Currently, dairy processors can incorporate technologies in their food safety efforts.

This chapter aims to review three different case studies dealing with dairy products to offer a summary of the variety of the latest innovative solutions to prolonging the SL of fresh dairy foods.

8.2 Fior di Latte Cheese

Fior di latte is a typical dairy product of Southern Italy, similar to mozzarella cheese. It is an unripened pasta filata cheese, manufactured either according to traditional procedures (raw milk inoculated with natural whey or milk cultures, raw milk ripened under special conditions, without starter addition) or by using pasteurized milk and commercial starter cultures of lactic acid bacteria. Although the cheese receives a heat treatment during curd stretching, postprocessing contamination by microorganisms may occur due to high moisture (from 55 % to 60 %) and high fat content (> 45 %), limiting its SL to a few days (Spano et al. 2003). To maintain the product quality, the cheese industry actually uses the brine, which is diluted whey or water, from the stretcher molder. During storage a mass transfer between the product and the liquid occurs that is related to the product composition, microbial growth, moisture degree, type of liquid, and storage conditions (Sinigaglia et al. 2008; Laurienzo et al. 2008).

Recently, attention has been focused on more attractive preservation methods based on the use of natural compounds with antimicrobial or antioxidant properties. Many spices and herbs may be used as natural alternatives to chemical additives (Tassau et al. 2000). The antimicrobial properties of plant essential oils are well established against a wide spectrum of microbes including fungi and bacteria (Deans and Ritchie 1987; Dorman and Deans 2000; Delaquis et al. 2002; Smith-Palmer et al. 2001; Paster et al. 1990). Altieri et al. (2005) successfully tested the effect of chitosan on mozzarella cheese. In that study, a lactic acid/chitosan solution was directly added to the starter used for mozzarella cheese manufacture to obtain a final concentration of 0.075 % low molecular weight chitosan (85 % deacetylation). Cheese samples, with and without chitosan, were stored at 4 °C for approximately 10 days. The monitoring of microbial populations demonstrated that chitosan inhibited the growth of some spoilage microorganisms, whereas it did not influence the growth of other microorganisms, such as *Micrococcaceae*, and lightly stimulated lactic acid bacteria. To calculate the product SL, *Pseudomonas* spp. and total coliforms were taken into account as target microbial groups for determining product unacceptability from a microbial point of view. According to Bishop and White (1986), a *Pseudomonas* sp. microbial load equal to 10^6 CFU/g represents the contamination level at which the alterations of the product start to appear. For total coliforms a threshold of 10^5 CFU/g was used (DPR 54/97; European Union 1997). Wherever the overall quality of a given product depends on several quality subindices (chemical, microbiological, or sensory), its SL is, by definition,

Table 8.1 Calculated microbial acceptability limit (MAL) (mean value along confidence interval) in relation to *Pseudomonas* spp. and total coliforms in mozzarella cheese with and without chitosan

	Sample	MAL *Pseudomonas* (day)	*P*-value[a]	MAL total coliforms (day)	*P*-value
M1[b]	Control sample	4.85 [3.64–5.85]	<0.0001	3.91 [3.20–4.54]	<0.0001
	Sample with chitosan	6.73 [6.17–7.49]		4.67 [3.92–5.37]	
M2	Control sample	3.97 [3.77–4.40]	0.1264	8.01 [7.41–8.48]	<0.0001
	Sample with chitosan	4.33 [3.56– 12.9]		9.25 [8.94–9.50]	

[a]Significance level of differences among means, referred to an unpaired student *t*-test with unequal variance
[b]M1 and M2 representing two independent mozzarella manufacturing processes

the time at which one of the product quality subindices reaches its threshold value. In the specific case of Altieri et al. (2005) the SL of the packaged mozzarella cheese was calculated on the basis of the microbiological assessment using the approach proposed by Corbo et al. (2006). It consisted in rearranging the Gompertz equation, as modified by Zwietering et al. (1990), in such a way that the microbial accept-ability limit (MAL) appeared directly as a parameter of the equation relating the log (cfu/g) to the storage time. The equation was reported in the modeling section in the Chap. 1, of this book; for the sake of simplicity it is also reported below:

$$\log\left(\frac{cfu}{g}\right) = \left[\log\left(\frac{cfu}{g}\right)\right]\max - A \cdot \exp\left\{-\exp\left\{\left[(\mu_{\max} \cdot 2.71) \cdot \frac{\lambda - MAL}{A}\right] + 1\right\}\right\}$$
$$+ A \cdot \exp\left\{-\exp\left\{\left[(\mu_{\max} \cdot 2.71) \cdot \frac{\lambda - t}{A}\right] + 1\right\}\right\},$$

(8.1)

where $\left[\log\left(\frac{cfu}{g}\right)\right]$ max is the decimal logarithm of the microbial acceptability limit, A is the maximum bacterial growth attained at the stationary phase, $\mu_{\max}$ is the maximal specific growth rate, λ is the lag time (day), and t is the time (day).

The results obtained by fitting Eq. 8.1 to the experimental data related to the evolution of *Pseudomonas* spp. and total coliforms (Table 8.1) highlight that chitosan slowed down the growth of coliforms during storage, leading to a slight increase in SL (approximately 1 day). To verify the repeatability of the results, even when working with different initial cell loads in the samples, different production batches were used (M1 and M2). Also in the case of pseudomonads, samples with chitosan showed a significant SL increase with respect to samples without chitosan (Table 8.1). According to an unstructured sensory analysis, no differences between chitosan and chitosan-free samples were recognized, demonstrating the validity of the method in prolonging cheese SL.

Table 8.2 Calculated microbial acceptability limit (MAL) (mean value along 95 % confidence interval) in relation to *Pseudomonas* spp. and total coliforms in mozzarella cheese

	Sample	MAL *Pseudomonas* (day)	MAL total coliforms (day)
	Control sample	1.58 [1.43–1.70]	1.77 [1.44–2.07]
Lemon	Active gel 500 ppm	2.06 [2.00–2.11]	2.23 [2.12–2.36]
	Active solution 500 ppm	2.45 [2.24–2.55]	2.61 [2.33–4.65]
	Control sample	1.24 [1.07–1.44]	1.78 [1.67–1.83]
Lemon	Active gel 1,000 ppm	2.22 [1.99–2.40]	2.15 [1.99–2.30]
	Active solution 1,000 ppm	3.12 [2.78–3.46]	2.03 [1.98–2.06]
	Control sample	1.37 [1.22–1.50]	1.77 [1.65–1.91]
Lemon	Active gel 1,500 ppm	2.20 [2.02–2.38]	1.82 [1.62–2.00]
	Active solution 1,500 ppm	2.99 [2.73–3.26]	2.20 [2.01–2.48]

Applications of natural antimicrobial agents to mozzarella packaging were also carried out. Conte et al. (2007) evaluated the effects of lemon extract, at three different concentrations, dissolved in the brine of mozzarella cheese and in a gel made up of sodium alginate, placed in the bottom of the tray. The cell load of both spoilage and dairy functional microorganisms were monitored at regular time intervals during storage. By fitting the experimental data of *Pseudomonas* spp. and total coliforms through the aforementioned Eq. 8.1, the microbial acceptability of the packaged dairy product (MAL) was calculated; data are reported in Table 8.2. The data show an increase in MAL values of all active packaged mozzarella cheese compared to the respective control sample, confirming that the investigated substance may exert an inhibitory effect on the microorganisms responsible for spoiling, without affecting the functional microbiota of the product.

Sinigaglia et al. (2008) also demonstrated that it is possible to prolong the SL of mozzarella by dissolving lysozyme and disodium ethylenediaminetetraacetic acid (Na_2-EDTA) in the packaging brine. Lysozyme is a lytic enzyme found in many natural systems and used in cheese manufacture to prevent the growth of lactate-fermenting and gas-forming *Clostridia* spp. (Crapisi et al. 1993). The antimicrobial spectrum of lysozyme could be enhanced when it is used with other substances, such as EDTA (Branen and Davidson 2004), disodium pyrophosphate, pentasodium tripolyphosphate (Boland et al. 2003), caffeic acid, and cinnamic acid (Masschalck and Michiels, 2003). Na_2-EDTA is not carcinogenic and does not show a bioaccumulation potential (SIAM 2001). Its NOAEL (no observed adverse lethal effect level) value is 500 mg $\cdot$ kg^{-1} $\cdot$ day^{-1}. It is approved for use in various foods at concentrations ranging from 35 to 800 ppm (Heimbach et al. 2000). In addition, it is approved for indirect food additives used as components of adhesive (21 CFR 175.105) (Code of Federal Regulations 1998) and as a component of an aqueous sanitizing solution used in food-processing equipment (21 CFR 178.1010) (Code of Federal Regulations 1998). Two experimental plans were carried out in the work of Sinigaglia et al. (2008): in the first phase, a dilute salt solution was used as the conditioning brine (with and without the antimicrobial mix), whereas in the second step, the mozzarella cheese was packaged in a diluted salt solution buffered to pH 6.5.

Table 8.3 Calculated microbial acceptability limit (MAL) (mean value along 95 % confidence interval) in relation to *Pseudomonas* spp. and total coliforms in fior di latte cheese

Sample	MAL *Pseudomonas* (day)	MAL total coliforms (day)
Control sample in diluted salt solution	3.62 [3.13–4.60]	3.74 [3.09–5.08]
Control sample in diluted salt solution acidified to pH 4.5	4.24 [3.59–4.96]	4.35 [4.09–5.87]
Sample in diluted salt solution with lysozyme[a] and Na$_2$-EDTA 50 mmol L^{-1}	6.59 [6.27–6.80]	–[b]
Sample in diluted salt solution with lysozyme[a] and Na$_2$-EDTA 20 mmol L^{-1}	5.65 [5.13–6.24]	–
Sample in diluted salt solution with lysozyme[a] and Na$_2$-EDTA 10 mmol L^{-1}	6.61 [6.16–7.12]	7.25 [6.62–8.8 × 10^{27}]

[a]0.25 mg mL^{-1}

[b]It was not possible to evaluate shelf life values because within 8 days the cell load of coliforms did not reach the breaking point

In the first step cheese was placed in individual polyethylene pouches containing 200 mL of the conditioning solution (12 % NaCl brine), which contained 0.25 mg · mL^{-1} of lysozyme (Sigma-Aldrich, Milan, Italy) and various concentrations of Na$_2$-EDTA (J.T. Baker, Milan, Italy), i.e., 50, 20, and 10 mmol · L^{-1}. In the second experimental phase, to prevent a decrease in pH due to the addition of Na$_2$-EDTA, cheese samples were packaged with 200 mL of a dilute salt solution to which a phosphate buffer (50 mmol · L^{-1}; pH 6.5; K$_2$HPO$_4$/KH$_2$PO$_4$: J.T. Baker) had been previously added. The conditioning solution contained lysozyme (0.25 mg · mL^{-1}) and Na$_2$-EDTA (50 and 20 mmol · L^{-1}). Mozzarella cheeses, packaged in buffered diluted brine, were used as the controls. As also reported in the literature, the microbial acceptability was evaluated through the population numbers of coliforms and Pseudomonadaceae by fitting Eq. 8.1 to the experimental data. The effect of lysozyme and Na$_2$-EDTA was similar for *Pseudomonas* spp. and total coliforms: an initial viability loss followed by an increase of cell number after 3 days of storage was observed in sample prepared with the diluted salt solution containing lysozyme and Na$_2$-EDTA 50 mmol · L^{-1}. Otherwise, microorganisms were able to proliferate in the controls and in the samples packaged in the conditioning brine that contained lower amounts of Na$_2$-EDTA. A similar result (decrease in the number of cells within 3 days, followed by an increase in the population) was recovered for the samples of the second phase. Table 8.3 reports the MAL values recorded in the first step of the study; similar results were also recorded in the samples of the second phase, with no significant differences between the two amounts of Na$_2$-EDTA or the two microbial target groups. As can be seen in Table 8.3, the addition of the active compounds to the conditioning solution prolonged the MAL evaluated through the coliforms. In fact, it was approximately 4 days for both the control samples, whereas for mozzarella packaged with higher concentration of Na$_2$-EDTA it was more than 8 days. A similar result was also obtained for *Pseudomonas* spp., thus justifying the assumption that the active compounds exerted an intrinsic antimicrobial action, unlike the acidification of the conditioning solution. As the microbial acceptability was evaluated through the cell numbers of two different targets, the microbial SL of mozzarella cheese was assumed to be the lowest value between them. In the conditions applied in this research, *Pseudomonas* spp. showed a MAL value lower than that observed for total coliforms, and the differences were significant ($p < 0.05$) for the sample with higher amounts of Na_2-EDTA, the product acceptability limit was approximately 6 days. Even though the effects recorded with the investigated antimicrobial compounds on product microbial quality seemed very interesting, sensory analyses need to be conducted because it can be supposed that the addition of Na_2-EDTA could affect the texture of mozzarella cheese, provoking sensory quality loss.

Gammariello et al. (2008) carried out a preliminary study to assess the efficiency of plant essential oils to be used as natural food preservatives in fior di latte cheese. In particular, *Salvia officinalis* and *Citrus limonum* from Alkott (Milan, Italy), *Thymus vulgaris* from Aboca (Arezzo, Italy), *Rosmarinus officinalis* from OTI [Domegge di Cadore (BL), Italy], *Citrus aurantium*, *Citrus sinensis*, and *Citrus paradisii* from Primavera [Lorenzana (PI), Italy], essential oils from *Citrus limonum*, *Melaleuca alternifolia*, *Mentha piperita*, *Propolis*, and *Vanilla planifolia* from Erbavita, citral 95 %, nonanoic acid 96 %, thymol, and limonene from Sigma-Aldrich (Italy) were directly dissolved into the fior di latte brine, at concentrations ranging from 500 to 10,000 ppm for sensorial testing. The active solutions were prepared 24 h prior to panel evaluation and were held at 10 °C in an odor-free environment. Immediately before testing, samples were equilibrated to room temperature. On the basis of the sensory compatibility between cheese and active compound, trained panelists selected a certain number of natural agents and discarded others due to the smell associated with the dairy product. Afterward, the selected active compounds were dissolved in the brine of packaged fior di latte samples at the desired concentrations. During storage at 10 °C for approximately 6 days the cell loads of spoilage (*Pseudomonas* spp. and total coliforms) and dairy useful microorganisms were monitored to calculate the MAL using the same modified version of the Gompertz equation (Eq. 8.1).

Results showed that most of the investigated essential oils did not prolong the microbial stability of mozzarella compared to the control sample; a few of them, such as lemon, sage, and thyme, extended the MAL values of the cheese, as reported in Table 8.4. The most interesting aspect of the work, which could promote the natural compounds' application in the dairy industry, is the fact that the preceding substances exerted an inhibitory effect on the microorganisms responsible for spoilage, without affecting the dairy microflora. The work also highlighted that not all essential oils act in the same manner. For this reason, the potential interactive effects of some natural compounds on the microbial stability of fior di latte cheese could be expected. As stated in the literature (Delaquis et al. 2002; Fu et al. 2007; Corbo et al. 2008; Mastromatteo et al. 2009), a combination of active compounds could lead to additive, synergistic, or antagonistic effects. Therefore, Gammariello et al. (2010) combined different active agents (sage and two lemon extracts) in the packaging brine of fior di latte cheese, according to a central composite design (CCD). The microbiological quality of the investigated fior di latte cheese, stored at 10 °C, was determined by monitoring the main spoilage

Table 8.4 Microbial acceptability limit (MAL) of fior di latte samples, assumed as lowest value between Mal^{Coliforms} and MAL^{Pseudomonas}

Sample	Manufacturing concern	MAL (day)
Control	Local market	$1.17 \pm 0.21^{a,b,c}$
Lemon 10,000 ppm	Spencer	$0.68 \pm 0.16^{a,b}$
Lemon 5,000 ppm	Spencer	0.58 ± 0.43^{a}
Citral 500 ppm	Sigma-Aldrich	$2.04 \pm 0.44^{b,c,d,e}$
Nonanoic acid 2,000 ppm	Sigma-Aldrich	NC
Grapefruit 5,000 ppm	Primavera	$1.65 \pm 0.10^{a,b,c,d,e}$
Lemon 5,000 ppm	Aboca	$1.60 \pm 0.21^{a,b,c,d,e}$
Lemon 5,000 ppm	Agrumigel	$0.82 \pm 0.40^{a,b}$
Lemon 1,500 ppm	Alkott	$1.39 \pm 0.10^{a,b,c,d}$
Lemon 3,000 ppm	Alkott	>6
Lemon 6,000 ppm	Alkott	>6
Lemon 1,500 ppm	Boyajian	$2.46 \pm 0.32^{c,d,e}$
Lemon 3,000 ppm	Boyajian	>6
Lemon 6,000 ppm	Boyajian	2.93 ± 0.39^{e}
Lemon 5,000 ppm	Erbavita	NC
Lemon 2,000 ppm	Esperis	$0.79 \pm 0.42^{a,b}$
Lemon 1,500 ppm	Primavera	$1.71 \pm 0.10^{a,b,c,d,e}$
Limonene 500 ppm	Sigma-Aldrich	0.55 ± 0.12^{a}
Orange 5,000 ppm	Erbavita	NC
Propolis 5,000 ppm	Erbavita	$0.73 \pm 1.57^{a,b}$
Rosemary 5,000 ppm	OTI	NC
Sage 1,500 ppm	Alkott	$2.04 \pm 0.39^{b,c,d,e}$
Sage 3,000 ppm	Alkott	>6
Sage 6,000 ppm	Alkott	>6
Sour orange 5,000 ppm	Primavera	$1.72 \pm 0.09^{a,b,c,d,e}$
Sweet orange 5,000 ppm	Primavera	$2.58 \pm 1.17^{d,e}$
Thyme 5,000 ppm	Aboca	$2.32 \pm 0.65^{c,d,e}$
Thymol 1,500 ppm	Sigma-Aldrich	$1.25 \pm 1.60^{a,b,c,d}$
Vanilla 5,000 ppm	Erbavita	NC

$^{a-e}$Means within the third column with different letters are significantly different ($p < 0.05$)
NC not calculated value

bacteria (*Pseudomonas* spp. and total coliforms). To quantitatively determine the effectiveness of the investigated antimicrobial compounds in slowing down the growth of selected spoilage microorganisms, the time at which the viable cell concentration reached its acceptability limit was calculated for each spoilage microbial group using the aforementioned Eq. 8.1. From Table 8.5 an increase in the MAL value was observed for all cheese samples packaged with the essential oils compared with the traditional product. Moreover, the CCD made it possible to evaluate the individual and interactive effects of sage and lemons on the microbial stability of fior di latte, thus making it possible to determine the optimal composition of natural compounds to be used for maximizing the microbiological acceptability without affecting sensory acceptance.

Table 8.5 Microbial acceptability limit (MAL) (days) of fior di latte samples, assumed as lowest value between $MAL^{Coliforms}$ and $MAL^{Pseudomonas}$

Sample	Lemon Alkott (ppm)	Lemon Boyajian (ppm)	Sage Alkott (ppm)	MAL (day)
Control	2,250	2,250	2,250	2.19 ± 0.28^a
Run 1	2,250	2,250	4,500	$3.99 \pm 1.10^{b,c,d}$
Run 2	2,250	4,500	2,250	$4.41 \pm 1.17^{b,c,d,e}$
Run 3	2,250	4,500	4,500	$3.53 \pm 0.86^{b,c}$
Run 4	4,500	2,250	2,250	$4.10 \pm 0.33^{b,c,d}$
Run 5	4,500	2,250	4,500	4.34 ± 0.76^{bcde}
Run 6	4,500	4,500	2,250	5.09 ± 0.39^{de}
Run 7	4,500	4,500	4,500	3.20 ± 0.70^{ab}
Run 8	3,000	3,000	3,000	3.90 ± 0.46^{bcd}
Run 9	3,000	3,000	1,500	4.78 ± 0.55^{cde}
Run 10	3,000	3,000	6,000	3.30 ± 0.64^{ab}
Run 11	3,000	1,500	3,000	3.68 ± 0.50^{bc}
Run 12	3,000	6,000	3,000	5.59 ± 0.65^{ef}
Run 13	1,500	3,000	3,000	4.34 ± 0.66^{bcde}
Run 14	6,000	3,000	3,000	5.24 ± 0.89^{de}
Run 15	3,000	3,000	3,000	4.75 ± 0.94^{cde}
Run 16	3,000	3,000	3,000	6.66 ± 0.68^f
Run 17	2,250	2,250	2,250	4.45 ± 0.23^{bcde}

Data presented $\pm$ standard deviation

[a–f]Means within the column with different letters are significantly different ($p < 0.05$)

Another important factor that limits fior di latte cheese distribution beyond the market borders is reflected in the transport costs due to the weight of the package. To overcome this limitation, some researchers proposed the substitution of brine with coating systems. The coatings are widely applied to extend the SL of minimally processed fruits and vegetables (Devlieghere et al. 2004; Rojas-Grau et al. 2007; Olivas et al. 2007). Kampf and Nussinovitch (2000) showed that bio-based polymeric layers, prepared by κ-carrageenan, alginate, and gellan gum, could reduce weight loss and improve the textural and sensory properties of semihard cheese. Park et al. (2004) reported that chitosan-based films with lysozyme incorporation enhanced the inhibition power of chitosan films on both Gram-positive and Gram-negative microorganisms, thus broadening their applications in ensuring food quality and safety of a low-moisture, part-skim mozzarella cheese that had been aged at least 60 days (Tillamook County Creamery Assoc., Tillamook, Oregon, USA) (Duan et al. 2007).

Sliced cheese, after inoculation, was subjected to three chitosan-lysozyme package applications: film, lamination on a multilayer coextruded film, and coating. Treated cheese received significant reductions in microbial growth. Incorporation of lysozyme in chitosan showed a greater antimicrobial effect than chitosan alone. Growth of mold was completely inhibited in cheese packaged with active films, while a slight reduction in mold populations was observed in cheese packaged with laminated films and coatings. The examples of coating applications to fresh cheese available in the literature are very scarce. Laurienzo et al. (2006) proposed the use of a gel to both substitute for the brine and increase product SL to buffalo mozzarella cheese. This dairy product is a typical pasta filata cheese from Southern Italy (so-called DOP mozzarella), with high moisture (55–62 %) and a high fat (greater than 45 %) content, characterized by a soft body and a juicy appearance and by a pleasant, fresh, sour, and slightly nutty flavor (Coppola et al. 1990). The product obtained from unpasteurized milk and natural whey can be stored, immersed in its mother solution, for 3 to 4 days at a temperature between 4 °C and 10 °C with no loss of its characteristics (Paonessa 2004), whereas the industrial product, made with pasteurized milk and selected starter, can maintain a prolonged SL, up to 20 days, but the taste is unsatisfactory when compared with artisanal mozzarella. The innovation proposed by Laurienzo et al. (2006) is based on the use of a gel, intended as a mix of agarose and gellan, to coat mozzarella by immersion at a temperature of approximately 40 °C. In the experimental plan, two temperatures (18 versus 4 °C) were used for the first 5 days of storage; after 5 days, all packages were stored in a refrigerator at 4 °C for up to 20 days. This is because some producers suggest keeping buffalo mozzarella at room temperature for the first 3 to 4 days. The results from microbiological analyses suggested that storage of water buffalo mozzarella cheese at room temperature was not advised because it promoted an increase in natural microflora, whereas the innovative gels established an unfavorable habitat for the growth of natural microflora and of enterococcal flora when the product was stored at refrigerated temperatures. From a mechanical point of view, after 5 days the control sample exhibited a strong diminution of compression strength, whereas the gel-stored sample maintained a very high strength (Fig. 8.1). Similar results were also recorded after 30 days of storage (Fig. 8.1). The developed packaging, without the need for chemical (citric acid or lactic acid) or thermal treatment of the milk, could guarantee a prolonged SL of traditional buffalo mozzarella of up to 20 days, with no influence on the values of pH, microbiological characteristics, and mechanical properties. Unfortunately, the analysis of the protein fraction from mozzarella cheese has not explained the differences in mechanical properties between gel-stored cheese and traditional product, probably due to the changes in content of Na and Ca between the two types of mozzarella cheese, following different osmotic exchanges of the different storage media (gel and water) (Laurienzo et al. 2008).

Conte et al. (2009) also successfully applied an active edible coating to fior di latte cheese, eliminating the brine and using MAP during packaging. To this end, each sample of fior di latte (50 g) was dipped into sodium alginate solution (8 % w/v) containing 0.25 mg/mL lysozyme and 50 mM of Na_2-EDTA; after coating, the samples were immersed into a 5 % (w/v) calcium chloride ($CaCl_2$) solution for 1 min. After drying, each sample was packaged in commercially available bags (thickness 95 μm, Valco, Bergamo, Italy) under ordinary and MAP conditions (30 % CO_2, 5 % O_2, and 65 % N_2). The gas composition of MAP was selected without reaching anaerobic conditions and considering that high carbon dioxide concentrations could provoke a bad aftertaste in the pack (Lichter et al. 2005). As controls, samples of fior di latte cheese without coating were also

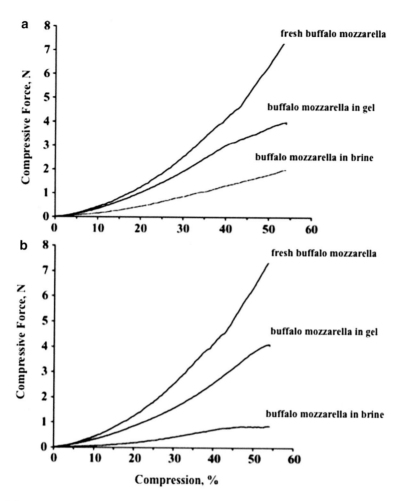

Fig. 8.1 Compression tests of water buffalo mozzarella stored for 5 days (**a**) and 30 days (**b**) at 4 °C in gel and in its own preservation liquid compared with fresh water buffalo mozzarella cheese

packaged in bags under MAP and in trays with traditional brine (2 % NaCl solution) and stored at 10 °C for 8 days. On various cheese samples determinations of pH, weight loss, headspace gas composition, microbial count, and sensory evaluation were carried out before packaging and after 1, 2, 3, 4, 7, and 8 days of storage. In terms of percentage weight loss, the most abundant amount of liquid was lost by the uncoated samples packaged in bags under MAP, due to the lack of brine. Conversely, the percentage weight loss of coated cheese was fairly similar to that of the control, suggesting that coating can work as a barrier to reduce weight loss, as happened with many fresh fruits and vegetables (Olivas et al. 2007; Rojas-Grau et al. 2007). From a microbiological point of view, significant differences were recorded between the samples. In particular, only the control overlapped the threshold value of *Pseudomonas* spp. (10^6 CFU/g); the other samples remained below this microbial limit during the entire observation period. For this reason, the MAL of all samples, except the control, was considered longer than the storage period (more than 8 days); whereas to quantitatively determine the microbial acceptability limit of the control packaging system, Eq. 8.1 was fitted to the experimental data. Similar results were also recorded for total coliforms (breaking point equal to 10^5 CFU/g). The better efficacy recorded with the coated samples can be ascribed to the active coating, and in particular to the direct contact of the antimicrobial compound to the food surface (Torres et al. 1985). Packaging under MAP without coating did not guarantee the same antimicrobial effect, but if combined to the coating, it further delayed the growth of spoilage microorganisms. The cell load of the flora type is not affected by the packaging system, according to other applications of MAP or lysozyme to dairy food (Eliot et al. 1998; Sinigaglia et al. 2008). From a sensory point of view, slight differences were recorded between samples. The values of sensory quality of the product differently packaged were highlighted by means of the following equation, which is a Gompertz equation as reparameterized by Corbo et al. (2006). The equation was reported in the modeling section in Chap. 1; for the sake of simplicity it is also reported below:

$$OSQ(t) = OSQ_{min} - A^Q \cdot \exp\left\{-\exp\left\{\left[(\mu^Q_{max} \cdot 2.71) \cdot \frac{\lambda^Q - SAL}{A^Q}\right] + 1\right\}\right\}$$
$$+ A^Q \cdot \exp\left\{-\exp\left\{\left[(\mu^Q_{max} \cdot 2.71) \cdot \frac{\lambda^Q - t}{A^Q}\right] + 1\right\}\right\}, \tag{8.2}$$

where $OSQ(t)$ is the fior di latte overall sensorial quality at time t, A^Q is related to the difference between the fior di latte overall sensorial quality attained at the stationary phase and the initial value of fior di latte overall sensorial quality, μ^Q_{max} is the maximal rate at which $OSQ(t)$ decreases, λ^Q is the lag time, OSQ_{min} is the fior di latte overall sensorial quality threshold value, SAL is the sensory acceptability limit (i.e., the time at which $OSQ(t)$ is equal to OSQ_{min}), and t is the storage time. The value of OSQ_{min} was set equal to 4. The sensory properties of the product did not exceed 3 days of storage, probably due to the weight loss that compromised its appearance. In the case under investigation, the SL of each tested sample was calculated as the lowest value between the MAL and SAL values (Table 8.6). SL prolongation was recorded for the samples packaged according to the techniques proposed by Conte et al. (2009), compared to the traditional system. It is worth noting that the unacceptability of fior di latte packaged in brine mainly depended on microbial quality, which limited the SL to just a few days. In contrast, MAP and active coatings could be advantageously proposed for industrial applications, due to the synergistic effects exerted on microbiological and sensory quality, and to the possibility to reduce costs related to product distribution.

Del Nobile et al. (2009b) combined the effects of chitosan previously tested by Altieri et al. (2005) to the packaging solutions proposed by Conte et al. (2009) to promote a further SL prolongation of fior di latte cheese. To this end, a chitosan

Table 8.6 Shelf life of fior di latte samples, assumed as lowest value among calculated microbial and sensorial acceptability limits

Sample	MAL (days)	SAL (days)	Shelf life (days)	Reference
Control sample	<1	3 ± 0	<1	Conte et al.
Control sample under MAP	<1	3 ± 1	<1	(2009)
Coated sample	>8	2 ± 1	2 ± 1^a	
Coated sample under MAP	>8	3 ± 1	3 ± 1^a	
Control sample	1.33 ± 0.17	6.75 ± 0.19	1.33 ± 0.17^b	Del Nobile et al.
Coated sample under MAP	5.03 ± 1.05	4.77 ± 0.92	4.77 ± 0.92^c	(2009b)
Sample with chitosan	1.36 ± 0.39	7.69 ± 0.23	1.36 ± 0.39^b	
Coated sample with chitosan under MAP	>8	5.05 ± 0.30	5.05 ± 0.30^c	

Data presented $\pm$ standard deviation
[a-c]Data in column with different small letters are significantly different ($p < 0.05$)
MAL microbiological acceptability limit calculated as the lowest value between $MAL^{Coliforms}$ and $MAL^{Pseudomonas}$, *SAL* sensorial acceptability limit for overall quality

solution containing lactic acid (1 %) was put into the working milk to obtain a final concentration of 0.012 % chitosan. On the packaged cheese stored at 4 °C, microbiological, pH, gas composition, and sensory changes were monitored over an 8-day period. To calculate the MAL and SAL values, Eqs. 8.1 and 8.2 were fitted to all the experimental data. Results highlighted that the proposed integrated approach based on the use of chitosan during the process, active coating prior to packaging and MAP during package sealing, was very effective in inhibiting the growth of the main spoilage microorganisms (*Pseudomonas* spp. and total coliforms). Environmental factors such as pH, water activity, temperature, and headspace gas composition can be considered individually as preservation methods when they are used at high doses, whereas according to hurdle technology they can be combined in a logical sequence to provide freshlike quality food products (Wiley 1994). Results also underlined that the control sample showed a very short SL limited to more or less 1 day, whereas the integrated approach developed in this study made it possible to obtain a significant SL prolongation to 5 days, most probably due to the synergistic effect between the active compounds and the atmospheric conditions in the package headspace. From a sensory point of view, the new packaging systems without brine compromised the product color, compared to the product stored in brine, whereas the presence of chitosan did not affect fior di latte sensory properties. Also in this case the SL was defined as the lowest value between $MAL^{coliforms}$, $MAL^{Pseudomonas}$, and SAL values (Table 8.6). It can be emphasized, on the basis of the data, that microbial quality limits the SL of lysozyme-free samples, whereas the sensory quality controls the SL of the coated samples, according to previous results in the literature (Conte et al. 2009). Chitosan alone did not prolong the SL of packaged cheese but, combined with other preservation strategies as an active coating and MAP, it can further enhance product acceptability to up to 5 days. Comparing this work to the work of Altieri et al. (2005) some differences can be highlighted. These differences could be ascribed to the type of chitosan (a low molecular weight chitosan in contrast to a high molecular weight chitosan) (Gerasimenko et al. 2004; Zheng and Zhu 2003; Tikhonov et al. 1996), to the different concentrations adopted (0.075 % versus 0.012 %), and to the different dissolving medium (whey in contrast to lactic acid).

8.3   Stracciatella Cheese

Stracciatella cheese is a typical Apulian dairy product, produced from cow's milk and packaged in rigid or flexible films of multilayer material, trays made of polyethylene/paper laminated films, and tetrapack-type packages (Robertson 1993). It is made up of a frayed curd mixed with a fresh cream that requires, like all fresh cheese, a refrigerated temperature during storage. It is also used to produce the burrata cheese, that consists of a ball of pasta filata cheese that contains stracciatella. Due to the product composition, i.e., very high moisture and fat content, the microbial spoilage represents the main factor responsible for product unacceptability. The literature does not report many studies on preservation of the product; the only works dealing with SL prolongation of stracciatella are two studies by Gammariello et al. (2009, 2011). The authors demonstrated the efficacy of MAP in preventing microbial proliferation, and in a subsequent step they combined the individuated proper headspace gas composition to the use of chitosan during the process, according to other experimental works also carried out on dairy foods that assessed the efficacy of chitosan as a valid preservative (Altieri et al. 2005; Del Nobile et al. 2009b). The potential of a suitable combination of carbon dioxide (CO_2), oxygen (O_2), and nitrogen (N_2) for extending the SL of dairy products has also been demonstrated in the literature, with effects that depend on the type of cheese, on the temperature, and on the barrier properties of the packaging materials (Alves et al. 1996; Eliot et al. 1998; Floros et al. 2000; Moir et al. 1993; Papaioannou et al. 2007; Mannheim and Soffer 1996). Gammariello et al. (2009) tested the effects of various combinations of $CO_2:N_2:O_2$ on the microbiological and sensory changes of stracciatella cheese stored at 8 °C for 8 days. To this end, a certain amount of cheese was packaged in a commercially available bag (thickness 95 μm, Valco, Bergamo, Italy) and sealed under different MAP conditions: M1 50/50 (CO_2/N_2), M2 95/5 (CO_2/N_2), M3 75/25 (CO_2/N_2), and M4 30/65/5 ($CO_2/N_2/O_2$). Cheeses in traditional tubs and under vacuum served as the controls. Determinations of microbial count, pH, headspace gas composition, and sensory quality were carried out before packaging and after 1, 2, 3, 4, 7, and 8 days of storage. To assess the microbial acceptability of the product, total coliforms and *Pseudomonas* spp. were selected as target spoilage microorganisms, and the MAL values were calculated using Eq. 8.1. The microbial counts of cheese samples packaged under MAP were lower than the controls. In particular, M2 and M3 were the most effective for inhibition, probably due to the higher concentration of CO_2 (Farber 1991; Gonzalez-Fandos et al. 2000). Counts of cocci and rod lactic acid bacteria in all stracciatella cheese samples remained unchanged, thus

Table 8.7 Shelf life of Stracciatella samples, assumed as lowest value among calculated microbial and sensorial acceptability limit

Sample	MAL (days)	SAL (days)	Shelf life (days)	Reference
Control sample	0.03 ± 3.23^a	2.29 ± 0.12^b	0.03 ± 3.23^a	Gammariello
Control sample under vacuum packaging	0.80 ± 1.19^a	2.17 ± 0.03^b	0.80 ± 1.19^a	et al. (2009)
Sample under MAP1	$1.13 \pm 0.33^{a,b,c}$	4.60 ± 0.51^a	1.13 ± 0.33^a	
Sample under MAP2	$1.37 \pm 0.24^{b,c}$	5.99 ± 0.36^c	1.37 ± 0.24^a	
Sample under MAP3	1.53 ± 0.16^c	4.45 ± 0.31^a	1.53 ± 0.16^a	
Sample under MAP4	0.81 ± 0.21^a	4.21 ± 0.15^a	0.81 ± 0.21^a	
Control sample	3.4 ± 0.2^a	4.3 ± 0.2^b	3.4 ± 0.2^a	Gammariello
Sample with chitosan I	4.0 ± 0.2^a	$5.2 \pm 0.3^{a,b}$	4.0 ± 0.2^a	et al. (2011)
Sample with chitosan II	3.9 ± 0.3^a	$4.9 \pm 0.2^{a,b}$	3.9 ± 0.3^a	
Sample with chitosan III	3.7 ± 0.2^a	$6.0 \pm 0.4^{a,c}$	3.7 ± 0.2^a	
Control sample under MAP	5.8 ± 0.3^b	7.5 ± 0.3^d	5.8 ± 0.3^b	
Sample with chitosan I under MAP	6.5 ± 0.2^c	5.8 ± 0.3^a	5.8 ± 0.3^b	
Sample with chitosan II under MAP	7.0 ± 0.6^c	$6.9 \pm 1.1^{c,d}$	6.9 ± 1.1^c	
Sample with chitosan III under MAP	6.8 ± 0.5^c	5.6 ± 1.0^a	5.6 ± 1.0^b	

Data presented $\pm$ standard deviation

[a-d]Data in column with different small letters are significantly different ($p < 0.05$)

MAL microbiological acceptability limit calculated as lowest value between $MAL^{Coliforms}$ and $MAL^{Pseudomonas}$, *SAL* sensorial acceptability limit for overall quality

MAP1 = 50:50 (CO_2:N_2); MAP2 = 95:5 (CO_2:N_2); MAP3 = 75:25 (CO_2:N_2); MAP4 = 30:65:5 (CO_2:N_2:O_2)

demonstrating that MAP conditions did not influence the growth of typical dairy microorganisms (Lioliou et al. 2001; Maniar et al. 1994). To calculate the product sensory acceptability, Eq. 8.2 was fitted to the sensory data. Significantly higher SAL values were recorded for the samples packaged under MAP, compared to the traditional packaging system and to the vacuum packaging solution. While in both the control samples the sensory properties were acceptable for approximately 2-day storage, the products under MAPs were acceptable for more than 4 days of storage. Combining the results of microbiological and sensory evaluations, the product SL can be inferred. From data reported in Table 8.7 the different MAP effects on stracciatella can be seen: M1 and M2 were the most effective MAPs to retain good sensory characteristics.

Considering the well-known antimicrobial properties of chitosan (Devlieghere et al. 2004; Hague et al. 2005; Kim et al. 2005; Yamada et al. 2005; Rabea et al. 2003) and the interesting effects exerted on other fresh dairy products (Altieri et al. 2005; Del Nobile et al. 2009b), Gammariello et al. (2011) combined the use of chitosan during the process of making stracciatella cheese with a packaging under MAP. To that end, a chitosan solution containing lactic acid (1 %) was put into the working milk to obtain a final concentration of 0.010 %, 0.015 %, and 0.020 % chitosan, respectively. All the samples with and without chitosan were packaged in

high-barrier bags in air and under MAP (75:25 CO_2:N_2). Determinations of
headspace gas composition, microbial count, pH, and sensory quality were carried
out before packaging and after 1, 4, 5, 6, 7, and 8 days of storage at 4 °C. The
$MAL^{Pseudomonas}$ and $MAL^{coliforms}$ values, obtained using Eq. 8.1, highlighted that
the presence of chitosan alone did not significantly affect the microorganisms,
whereas chitosan works in synergy with the MAP to inhibit microbial growth,
according to previously results recorded on fior di latte cheese (Del Nobile et al.
2009b). The pH values were not affected by the integrated approach, suggesting
that the observed antimicrobial activity had to be exclusively ascribed to the
effectiveness of the combined strategy. For the sensory quality the same mathematical approach (Eq. 2 of part I), previously used in other works (Conte et al. 2009;
Del Nobile et al. 2009b; Gammariello et al. 2009), was also adopted. As expected,
the quality of the tested cheese steadily decreased, regardless previously of the
packaging strategy. In particular, chitosan alone improved the odor of samples,
compared to the control, and the effects were further enhanced when combined to
the MAP conditions. The SL data, intended as the lowest value between MAL and
SAL, highlighted that a certain improvement could be obtained when chitosan was
combined with MAP because this technology controlled microbial proliferation
without affecting sensory quality (Table 8.7).

8.4 Ricotta Cheese

Ricotta is a whey cheese of the Mediterranean area, prepared traditionally by
heating whey from sheep or goat milk in open kettles. It is very similar in
composition to cottage cheese. After coagulation, the curd mass floats to the surface
and is scooped off and placed in perforated trays for drainage. Due to the high
moisture content, the initial pH above 6, and the low salt content, ricotta is very
susceptible to spoilage by fungi and bacteria (Fleet and Mian 1987; Pintado et al.
2001). Hough et al. (1999) reported SL values of 33, 13.5, and 5.5 days for
commercial ricotta cheese packaged in polyethylene bags and kept at 6, 17, and
25 °C, respectively. Other than refrigerated temperatures, little data are available in
the literature dealing with preservation techniques of ricotta. Vacuum packaging
(Tsiotsias et al. 2002) and antimicrobial compounds (Samelis et al. 2003) were both
applied to Greek dairy products obtained in a way similar to ricotta. MAP was also
applied to semihard and fresh cheese with different results that reflected the
different composition of the food product, the initial microbial contamination, the
postprocess hygienic conditions, and the storage temperature (Colchin et al. 2001;
Gonzalez-Fandos et al. 2000; Piergiovanni et al. 1993; Juric et al. 2003; Conte et al.
2009; Del Nobile et al. 2009b; Gammariello et al. 2011). Generally, high levels of
CO_2 were used, due to the antimicrobial effects of carbon dioxide on fungi and
bacteria, in combination with N_2 to avoid the collapse of the package (Mannheim
and Soffer 1996; Pintado and Malcata 2000). Papaioannou et al. (2007) proved that
the use of MAP (70:30 CO_2:N_2) extended the SL of typical Greek whey cheese by

approximately 20 days, compared to another gas mixture with lower CO_2 concentration. Cheese containing high moisture and fat have been preserved with a middle concentrations of CO_2 levels. Gonzalez-Fandos et al. (2000) studied five different $CO_2:N_2$ mixtures (from 20 % to 100 % CO_2) and demonstrated that the most effective gas combinations for extending the SL of a typical fresh cheese from Spain and retaining good sensory characteristics are based on CO_2 content ranging between 50 % and 60 %. A new contribution to the dairy field was provided by the study of Del Nobile et al. (2009a) that tested the application of different MAP conditions to ricotta cheese stored at 4 °C for 8 days. Ricotta from sheep milk was portioned, packaged in high-barrier, multilayer plastic bags, and sealed under three different MAPs: 50:50 ($CO_2:N_2$) (MAP50), 70:30 ($CO_2:N_2$) (MAP70), and 95:5 ($CO_2:N_2$) (MAP95). The quality loss of the product was assessed by monitoring microbial and physicochemical parameters. Tested MAP conditions did not seem to influence, to a great extent, the growth cycle of lactic acid bacteria, probably due to their facultative anaerobic nature and the high value of moisture content and pH of the whey cheese that promote lactic acid bacteria proliferation (Papaioannou et al. 2007; Pintado et al. 2001). Considering that spoilage generally starts to occur at 10^7 CFU/g of total viable count (ICMSF 1984), this level was set as a threshold to assess the microbial acceptability. The fitting procedure allowed for calculating sample SLs: 1.14 ± 0.34 for the control, 0.55 ± 0.5 for MAP50, 0.36 ± 0.6 for MAP70, and 3.37 ± 0.7 for MAP95. Substantial differences between samples were also recorded in terms of yellow color parameter [b*] that remained fairly constant over time for samples stored under MAP compared to the control. To sum up, MAP, when compared with air, retained product quality better; in particular, a gas mix richer in CO_2 inhibited microbial growth without significantly affecting lactic acid bacteria and also promoting a better preservation of color characteristics.

References

Albenzio M, Corbo MR, Rehman SU, Fox PF, De Angelis M, Corsetti A, Sevi A, Gobbetti M (2001) Microbiological and biochemical characteristics of Canestrato Pugliese cheese made from raw milk, pasteurized milk or by heating the curd in hot whey. Int J Food Microbiol 67:35–48

Albenzio M, Marino R, Caroprese M, Santillo A, Annicchiarico G, Sevi A (2004) Quality of milk and of Canestrato Pugliese cheese from ewes exposed to different ventilation regimens. J Dairy Res 71:434–443

Altieri C, Scrocco C, Sinigaglia M, Del Nobile MA (2005) Use of chitosan in prolonging mozzarella cheese shelf life. J Dairy Sci 88:2683–2688

Alves RMV, Sarantopoulos CIG, Van Dender AGF, Faria JAF (1996) Stability of sliced mozzarella cheese in modified atmosphere packaging. J Food Prot 59:838–844

Bishop JR, White CH (1986) Assessment of dairy product quality and potential shelf-life: a review. J Food Prot 49:739–753

Boland JS, Davidson PM, Weiss J (2003) Enhanced inhibition of Escherichia coli O157: H7 by lysozyme and chelators. J Food Prot 66:1783

Branen JK, Davidson PM (2004) Enhancement of nisin, lysozyme, and monolaurin antimicrobial activities by ethylenediaminetetraacetic acid and lactoferrin. Int J Food Microbiol 90:63–74

Byrne RD, Bishop JR (1998) Control of microorganisms in dairy processing: dairy product safety systems. In: Marth E, Steele J (eds) Applied dairy microbiology. Marcel Dekker, New York, pp 405–430

Code of Federal Regulations (1998) Title 21: food and drugs. US Government Printing Office, Washington, DC

Colchin LM, Owens SL, Lyubachevskaya G, Boyle-Roden E, Russek-Cohen E, Rankin SA (2001) Modified atmosphere packaged cheddar cheese shreds: influence of fluorescent light exposure and gas type on color and production of volatile compounds. J Agric Food Chem 49:2277–2282

Conte A, Scrocco C, Sinigaglia M, Del Nobile MA (2007) Innovative active packaging system to prolong the shelf life of mozzarella cheese. J Dairy Sci 90:2126–2131

Conte A, Gammariello D, Di Giulio S, Attanasio M, Del Nobile MA (2009) Active coating and modified atmosphere packaging to extend the shelf life of Fior di latte cheese. J Dairy Sci 92:887–894

Coppola S, Villani F, Coppola R, Parente E (1990) Comparison of different starter systems for water-buffalo mozzarella cheese manufacture. Lait 70:411–423

Corbo MR, Del Nobile MA, Sinigaglia M (2006) A novel approach for calculating shelf-life of minimally processed vegetables. Int J Food Microbiol 106:69–73

Corbo MR, Speranza B, Filippone A, Granatiero S, Conte A, Sinigaglia M, Del Nobile MA (2008) Study on the synergic effect of natural compounds on the microbial quality decay of packaged fish hamburger. Int J Food Microbiol 127:261–267

Crapisi A, Lante A, Pasini G, Spettoli P (1993) Enhanced microbial cell lysis by the use of lysozyme immobilized on different carrier. Process Biochem 28:17–21

Deans SG, Ritchie G (1987) Antibacterial properties of plant essential oils. Int J Food Microbiol 5:165–180

Del Nobile MA, Conte A, Incoronato AL, Panza O (2009a) Modified atmosphere packaging to improve the microbial stability of ricotta. Afr J Food Microbiol 3:137–142

Del Nobile MA, Gammariello D, Conte A, Attanasio M (2009b) A combination of chitosan, coating and modified atmosphere packaging for Fior di Latte cheese. Carbohydr Polym 78:151–156

Delaquis PJ, Stanich K, Girad B, Mazza G (2002) Antimicrobial activity of individual and mixed fractions of dill, cilantro, coriander and eucalyptus essential oils. Int J Food Microbiol 74:101–109

Devlieghere F, Vermeulen A, Debevere J (2004) Chitosan: antimicrobial activity, interactions with food components and applicability as a coating on fruit and vegetables. Food Microbiol 21:703–714

Dorman HJD, Deans SG (2000) Antimicrobial agents from plants: antibacterial activity of plant volatile oils. J Appl Microbiol 88:308–316

Duan J, Park SI, Daeschel MA, Zhao Y (2007) Antimicrobial chitosa-lysozyme films and coatings for enhancing microbial safety of mozzarella cheese. J Food Sci 72:355–362

Eliot SC, Vuillemard JC, Emond JP (1998) Stability of shredded mozzarella cheese under modified atmospheres. J Food Sci 63:1075–1080

European Union (1997) DPR 54/97. Regolamento recante attuazione delle Dir. 92/46 e 92/47/CEE in materia di produzione e immissione sul mercato di latte e di prodotti a base di latte, Brussels

Farber JM (1991) Microbiological aspects of modified atmosphere packaging technology-a review. J Food Prot 54:58–70

Fleet GH, Mian MA (1987) The occurrence and growth of yeasts in dairy products. Int J Food Microbiol 4:145–155

Floros JD, Nielsen PV, Farkas JK (2000) Advances in modified atmosphere and active packaging with applications in the dairy industry. Packaging of milk products, bulletin of the IDF, 346. Brussels, International Dairy Federation, pp 22–28

Fu YJ, Zu YG, Chen LY, Shi XG, Wang Z, Sun S, Efferth T (2007) Antimicrobial activity of clove and rosemary essential oils alone and in combination. Phytother Res 21:989–994

Gammariello D, Di Giulio S, Conte A, Del Nobile MA (2008) Effects of natural compounds on microbial safety and sensory quality of Fior di Latte cheese, a typical Italian cheese. J Dairy Sci 91:4138–4146

Gammariello D, Conte A, Di Giulio S, Attanasio M, Del Nobile MA (2009) Shelf life of Straciatella cheese under modified atmosphere packaging. J Dairy Sci 92:483–490

Gammariello D, Conte A, Attanasio M, Del Nobile MA (2010) Study on the combined effects of essential oils on microbiological quality of Fior di Latte cheese. J Dairy Res 77:144–150

Gammariello D, Conte A, Attanasio M, Del Nobile MA (2011) A study on the synergy of modified atmosphere packaging and chitosan on stracciatella shelf life. J Food Process Eng 34:1394–1407

Gerasimenko DV, Avdienko ID, Bannikova GE, Zueva OY, Varlamov VP (2004) Antibacterial effects of water-soluble low molecular-weight chitosans on different microorganisms. Appl Biochem Microbiol 40:253–257

Gonzalez-Fandos E, Sanz S, Olarte C (2000) Microbiological, physicochemical and sensory characteristics of Cameros cheese packaged under modified atmospheres. Food Microbiol 17:407–414

Hague T, Chen H, Ouyang W, Martoni C, Lawuyi B, Urbanska AM (2005) Superior cell delivery features of poly (ethylene glycol) incorporated alginate, chitosan and poly-L-lysine microcapsules. Mol Pharm 21:29–36

Heimbach J, Rieth S, Mohamedshah F, Slesinski R, Samuel-Fernando P, Sheenan T et al (2000) Safety assessment of iron EDTA [sodium iron (Fe3+) ethylene diamine tetraacetic acid]: summary of toxicological, fortification and exposure data. Food Chem Toxicol 38:99–111

Hough G, Puglieso ML, Sanchez R, Mendez da Silva O (1999) Sensory and microbiological shelf life of a commercial Ricotta. J Dairy Sci 82:454–459

ICMSF (International Commission on Microbiological Specifications for Foods) (1984). Microorganismos de los alimentos. 1 Tecnicas de analisi microbiologica. Ed. Acribia, Zaragoza.

Juric M, Bertelsen G, Mortensen G, Petersen MA (2003) Light-induced color and aroma changes in sliced modified atmosphere packaged semi-hard cheeses. Int Dairy J 13:239–249

Kampf N, Nussinovitch A (2000) Hydrocolloid coating of cheeses. Food Hydrocolloid 14:531–537

Kim HJ, Chen F, Wang X, Rajapakse NC (2005) Effect of chitosan on the biological properties of sweet basil (Ocimum basilicum L.). J Agric Food Chem 53:3696–3701

Laurienzo P, Malinconico M, Pizzano R, Manzo C, Piciocchi N, Sorrentino A, Volpe MG (2006) Natural polysaccharide-based gels for dairy food preservation. J Dairy Sci 89:2856–2864

Laurienzo P, Malinconico M, Mazzarella G, Petitto F, Piciocchi N, Stefanile R, Volpe MG (2008) Water buffalo mozzarella cheese stored in polysaccharide-base gels: correlation between prolongation of the shelf life and physicochemical parameters. J Dairy Sci 91:1317–1324

Lichter A, Zutahy Y, Kaplunov T, Aharoni N, Lurie S (2005) The effect of ethanol dip and modified atmosphere on prevention of Botrytis rot of table grapes. Hortic Technol 15:284–291

Lioliou K, Litopoulou-Tzanetaki E, Tzanetakis N, Robinson RK (2001) Changes in the microflora of Manovri, a traditional Greek whey cheese, during storage. Int J Dairy Technol 54:100–106

Maniar AB, Marcy JE, Russell-Bishop J, Duncan SE (1994) Modified atmosphere packaging to maintain direct-set cottage cheese quality. J Food Sci 59:1305–1308

Mannheim CH, Soffer T (1996) Shelf-life extension of cottage cheese by modified atmosphere packaging. Lebensmittel-Wissenschaft und-Tecnologie 29:767–771

Masschalck B, Michiels CW (2003) Antimicrobial properties of lysozyme in relation to foodborne vegetative bacteria. Crit Rev Microbiol 29:191–214

Mastromatteo M, Lucera A, Sinigaglia M, Corbo MR (2009) Combined effects of thymol, carvacrol and temperature on the quality of non-conventional poultry patties. Meat Sci 83:246–254

Moir CJ, Eyles MJ, Davey JA (1993) Inhibition of pseudomonas in cottage cheese by packaging in atmospheres containing carbon dioxide. Food Microbiol 10:345–351

O'Reilly CE, Murphy PM, Kelly AL, Guinee TP, Auty MAE, Beresford TP (2002) The effect of high pressure treatment on the functional and rheological properties of mozzarella cheese. Innov Food Sci Emerg Technol 3:3–9

Olivas GI, Mattinson DS, Barbosa-Canovas GV (2007) Alginate coatings for preservation of minimally processed 'Gala' apples. Postharvest Biol Technol 45:89–96

Paonessa A (2004) Influence of the preservation liquid of Mozzarella di Bufala Campana D.O.P. on some aspects of its preservation. Bubalus bubalis IV:30–36

Papaioannou G, Chouliara I, Karatapanis AE, Kontominas MG, Savvaidis IN (2007) Shelf-life of a Greek whey cheese under modified atmosphere packaging. Int Dairy J 17:358–364

Park S-I, Daeschel MA, Zhao Y (2004) Functional properties of antimicrobial lysozyme-chitosan composite films. J Food Sci 69:215–221

Paster N, Juven BJ, Shaaya E, Menasherov M, Nitzan R, Weisslowicz H, Ravid U (1990) Inhibitory effect of oregano and thyme essential oils on moulds and food-borne bacteria. Lett Appl Microbiol 11:33–37

Piergiovanni L, Fava P, Moro M (1993) Shelf life extension of taleggio cheese by modified atmosphere packaging. Int J Food Sci 5:115–119

Pintado ME, Malcata FX (2000) The effect of modified atmosphere packaging on the microbial ecology in Requeijao a Portuguese whey cheese. J Food Process Preserv 24:07–124

Pintado ME, Macedo AC, Malcata FX (2001) Review: technology, chemistry and microbiology of whey cheeses. Int J Food Sci Technol 7:105–116

Rabea EI, Badawy MET, Stevens CV, Smagghe G, Steurbaut W (2003) Chitosan as antimicrobial agent: applications and mode of action. Biomacromolecules 4:1457–1465

Rojas-Grau MA, Raybaudi-Massilia RM, Soliva-Fortuny RC, Avena-Bustillos RJ, McHugh TH, Martin-Belloso O (2007) Apple puree-alginate edible coating as carrier of antimicrobial agents to prolong shelf-life of fresh-cut apples. Postharvest Biol Technol 45:254–264

SIAM (2001) SIDS initial assessment profile. CAS No. 60-00-4. EDTA. http://www.jetoc.or.sp/ HP-SIDSpdffiles/60-00-4.pdfwww.jetoc.or.sp/HP-SIDSpdffiles/60-00-4.pdfS.   Accessed 19 July.07

Sinigaglia M, Bevilacqua A, Corbo MR, Pati S, Del Nobile MA (2008) Use of active compounds for prolonging the shelf life of mozzarella cheese. Int Dairy J 18:624–630

Smith-Palmer A, Stewart J, Fyfe L (2001) The potential application of plant essential oils as natural food preservatives in soft cheese. Food Microbiol 18:463–470

Spano G, Goffredo E, Beneduce L, Tarantino D, Dupuy A, Massa S (2003) Fate of *Escherichia coli* O157: H7 during the manufacture of mozzarella cheese. Lett Appl Microbiol 36:73–76

Subramaniam PJ (1993) Miscellaneous applications. In: Prry RT (ed) Principles and applications of modified atmosphere packaging of foods. Blackie Academic & Professional, London, pp 170–188

Tassau C, Koutsoumanis K, Nychas GJE (2000) Inhibition of Salmonella enteritidis and Staphylococcus aureus in nutrient broth by mint essential oil. Food Res 33:273–280

Tikhonov VE, Radigina LA, Yamskov IA (1996) Metal-chelating chitin derivatives via reaction of chitosan with nitrilotriacetic acid. Carbohydr Res 290:33–41

Torres JA, Bouzas JO, Karel M (1985) Microbial stabilization of intermediate moisture food surfaces: II. Control of surface pH. J Food Process Preserv 9:93–106

Tsiotsias A, Savvaidis I, Vassila A, Kontominas MG, Kotzekidou P (2002) Control of Listeria monocytogenes by low-dose irradiation in combination with refrigeration in the soft whey cheese Anthotyros. Food Microbiol 19:117–126

Wiley RC (1994) Preservation methods for minimally processed refrigerated fruits and vegetables. In: Wiley RC (ed) Minimally processed refrigerated fruits and vegetables. Chapman & Hall, New York, pp 66–134

Yamada K, Akiba Y, Shibuya T, Kashiwada A, Matsuda K, Hirata M (2005) Water purification through bioconversion of phenol compounds by tyrosinase and chemical adsorption by chitosan beads. Biotechnol Prog 21:823–829

Zheng LY, Zhu JF (2003) Study on antimicrobial activity of chitosan with different molecular weight. Carbohydr Polym 54:527–530

Zwietering MH, Jogenburger I, Rombouts FM, van't Riet K (1990) Modeling of the bacterial growth curve. Appl Environ Microbiol 56:1875–1881

Chapter 9
Packaging for the Preservation of Meat- and Fish-Based Products

9.1 Introduction

Fresh meat and fish products are distributed under chilled conditions, while they are storable for a long time, they are sold frozen. Bacterial spoilage is the main factor responsible for meat deterioration. The high moisture content, high nutrient content, neutral pH, and presence of glycogen as a microbial substrate are all causes that work in concert to produce deterioration through the action of aerobic and anaerobic spoilage species (Ingram and Dainty 1971). Color change of red meat, as a function of the oxygen available on the product surface, also indicates product deterioration. As a consequence of the shrinking of myofibrils, a certain amount of fluid exudation, known as weep, appears in the package, thus compromising visual quality. The phenomenon is generally controlled by using film that is not too tightly shrink-wrapped and by placing an absorbent pad on the bottom of the package. For this reason, the use of vacuum packaging, although useful to control oxidative rancidity and prevent microbial growth, increases weep. Nonshrinkable film or thermoformed trays of polyethylene or polyvinyl chloride are generally used to better control microbial proliferation, even if in these conditions oxidative discoloration is accelerated. A large amount of information is also available on the packaging of meat under modified atmosphere packaging (MAP) and through the use of active and intelligent packaging (Kerry et al. 2006). However, the optimal composition of headspace for fresh meat and meat-derived products is not a simple task to accomplish (McMillin 2008).

Fish can be considered more perishable than meat due to the rapid autolysis of the enzymes and the oxidation of unsaturated oils (Gram and Dalgaard 2002). Microbial spoilage also occurs, with a consequent breakdown of proteins to produce volatile compounds responsible for unpleasant odors. The principal damage following a microbial contamination in fish can be ascribed to the formation of biogenic amines and spoilage compounds such as trimethylamine. The biogenic

M.A. Del Nobile and A. Conte, *Packaging for Food Preservation*,
Food Engineering Series, DOI 10.1007/978-1-4614-7684-9_9,
© Springer Science+Business Media New York 2013

amines are the result of unspecified decarboxylic reactions of amino acids by the action of microorganisms usually present in foods. They are involved in many physiological actions, such as the regulation of body temperature, stomach volume, and pH, but if they are present in high concentrations they can be toxic to the organism. Istamin is very toxic; the other biogenic amines, like putrescin, cadaverin, and spermidin, in certain concentrations, increase the effect of istamin. The formation of trimethylamine imparts the typical odor of rotten fish (González-Fandos et al. 2005). It is the result of the reduction in the trimethylamine oxide (TMAO) present in fish tissue, governed by *Alteromonas putrefaciens*. TMAO is a waste product derived from ammonia and from trimethylated compound degradations that is saved in tissue like oxide. Traditional packages of chilled fish are boxes with ice. For more delicate fish, such as fillets, plastic trays wrapped with film are used. The most suitable and extensively reviewed technology for perishable seafood product is MAP, but the selection of the proper gas combination is important to attain the desired effects (Mastromatteo et al. 2010; Sivertsvik et al. 2002). Vacuum packaging can also better retain fish quality compared to air, but the effectiveness depends on the raw materials and storage conditions. Active and intelligent packaging systems have also been tried; however, in the context of the so-called hurdle concept, the most effective strategy to increase the product shelf life (SL) is a combination of more than one preservation technology (Belcher 2006; Corbo et al. 2009a; Leistner 2000).

This chapter deals with two specific meat- and fish-based products: (1) minced meat in the forms of patties, meatballs, and hamburger and (2) fish burgers. The most widespread preservation methods, with a particular emphasis on the role of the packaging to prolong product SL, are presented and discussed in detail.

9.2 Fresh Minced Meat

Minced meat is a rich nutrient matrix that provides a suitable environment for the proliferation of meat spoilage microorganisms and common foodborne pathogens. In fact, fresh meat is one of the most perishable foods in commerce. Many factors influence meat SL, such as pH, water content, oxygen availability, and food composition. During storage these factors could promote spoilage bacterial growth and oxidative processes, which, in turn, provoke meat deterioration in flavor, texture, and color (Ingram and Dainty 1971). In particular, protein oxidation may decrease eating quality by reducing tenderness and juiciness and enhancing flavor deterioration and discoloration. Among the changes in proteins caused by oxidation is the formation of hydroperoxides and carbonyls, inter- and intramolecular cross-linking through the formation of disulfide bonds and dityrosine, fragmentation of the peptide backbone, and decreased protein solubility (Xiong 2000). On the other hand, lipid peroxidation leads to the formation of prooxidant substances capable of reacting with oxymyoglobin and resulting in metamyoglobin formation, thus compromising meat color (Yin and Faustman 1993).

Considering that food safety is a top priority for authorities and consumers worldwide, the hygienic condition of minced meat, in the form of hamburgers, meatballs, or patties, is of continuing concern in view of the association between enteric diseases, such as those caused by pathogenic strains of *Escherichia coli*, and hamburger consumption. Obviously, the microbiological condition of fresh meat-based products will be largely affected, if not wholly determined, by the microbiological condition of the manufacturing beef from which the products are prepared. It is now widely recognized that the hygienic conditions of raw meats can be assured only by the development of hazard analysis critical control point (HACCP) systems for meat production processes (USDA 1995). Nevertheless, the prevalence of foodborne pathogens and the reported number of cases and outbreaks is still high, affecting personal lives, business, and national economies (Aymerich et al. 2008). Adequate preservation technologies must be applied in order to preserve meat safety and quality. Several strategies have been used to improve the quality of fresh meat, such as the hygienic conditions of animals at slaughter, the control of spread of contamination during meat handling, the application of nonthermal preservation technologies, and the improvement of storage conditions in terms of temperature, light exposure, gas composition of package headspace, and active packaging (Aymerich et al. 2008; Mastromatteo et al. 2010; McMillin 2008).

Aymerich et al. (2008) presented an extensive overview dealing with the numerous preservation technologies to be applied to fresh meat, divided into nonthermal alternative preservation methods and thermal alternative technologies. In the first group, the high hydrostatic pressure (HHP) treatment, also known as ultra-high-pressure (UHP) treatments or high-pressure processing (HPP), irradiation, light pulses, natural biopreservatives, together with active packaging, have been proposed and discussed. These techniques are efficient at inactivating the vegetative microorganisms most commonly related to foodborne diseases, but not spores. Irradiation and HHP offer the possibility of processing packaged meat products, but both of them require a high investment and entail high maintenance costs. Moreover, the microbial resistance and the role of bacterial stress must be addressed in order to optimize treatments and support these technologies from a legislative point of view (Hugas et al. 2002; Gola et al. 2000; Sommers and Boyd 2006).

The use of natural or controlled microflora, mainly represented by lactic acid bacteria (LAB) or their antimicrobial products such as lactic acid, bacteriocins, and others, is a more natural technique for meat preservation (Cannarsi et al. 2008; Jacobsen et al. 2003; Pawar et al. 2000). Biopreservation offers an indirect way to apply antimicrobials and seems to be the most accepted way for consumers and producers. In minced meat, the bioprotective effect of several LAB and their bacteriocins have been shown to be effective antilisterial agents (Hugas 1998; Hugas et al. 1998; Schillinger et al. 1991). Although LAB are recognized as safe and valid SL extenders in a great variety of meat products, few cultures have been introduced to the market as starters or bioprotective cultures with the aim of contributing to microbiological safety (Aymerich et al. 2008). Today nisin is the only bacteriocin commercially available and accepted in the positive list of food additives (E234), although its use in meat products is not regulated (Bernbom et al. 2006).

The increasing consumer demand for healthier meat products, free of chemical additives, has also greatly promoted a search for other natural additives besides LAB that are of animal or plant origin, characterized by antioxidant and antimicrobial properties, and suitable for a wide spectrum of meat applications. One of the most studied categories is represented by the essential oils. The antimicrobial properties of essential oils are well documented in liquid media, in model structured systems, and in in vivo applications dealing with minced meat by direct addition to food (Ahn and Nam 2004; Fernández-López et al. 2005; Jo et al. 2003; Mitsumoto et al. 2005; Nissen et al. 2004; Racanicci et al. 2004; Skandamis et al. 2000). The effects of plant phenolics on protein and lipid oxidation (Dzudie et al. 2004; Vuorela et al. 2005) and on meat products fortified with n-3 fatty acids are also documented (Fernández-Ginés et al. 2005; Lee et al. 2006a, b). The effects of chitosan in meat preservation have been studied as well. Sagoo et al. (2002) showed that treatment with chitosan increased the SL of raw sausages stored at chilled temperatures from 7 to 15 days. These authors concluded that the antilisterial action of chitosan was particularly notable. This finding may promote chitosan as a valid substitute of the commonly used sulphites, which have a long history of safe use in meat products but whose application is linked to the aggravation of asthmatic and other respiratory problems in some sensitive individuals. Darmadji and Izumimoto (1994) showed that 0.01 % of chitosan inhibited the growth of some spoilage bacteria such as *Bacillus subtilis*, *E. coli*, *Pseudomonas fragi*, and *S. aureus* in meat patties. During incubation at 30 °C for 48 h or storage at 4 °C for 10 days, 0.5–1 % chitosan inhibited the growth of spoilage bacteria, reduced lipid oxidation, and resulted in better sensory attributes.

Extension of meat SL with MAP alone requires matching product and packaging material through careful selection, proper gas mixes, online analysis of the packaged products, detection of leaking packages, and offline testing for overall quality control (Stahl 2007). Common gases used in MAP are carbon dioxide (to inhibit bacterial growth), oxygen (to prevent anaerobic growth and to retain meat color), and nitrogen (to avoid oxidation of fats and pack collapse). These gases can be applied individually or in combination to achieve optimum effect, depending on the specific requirements of the particular food being preserved (Ellis et al. 2006; Jacobsen and Bertelsen 2002). The properties of meat packaged under MAP have been widely discussed in numerous reviews, and although voluminous information is available in the literature, further research is still essential to define the optimal headspace composition for each meat-based product. The extensive review of McMillin (2008) represents an excellent effort to highlight the actual and future potential of MAP for meat. The author suggested that the success and continuation of the many different retail MAP formats depend on product, package, and system interactions, the relationships of processors and retailers, and consumer acceptance of the merchandising format. Advances in plastic materials and equipment have propelled advances in MAP, but other technological and logistical considerations

Table 9.1 Overview of studies testing the combined use of MAP and natural compounds in minced meat in forms of patties, hamburgers, and meatballs

Food	Natural compounds	Packaging conditions	Sources
Minced beef	Exogenous α-tocopherol, dispersed in olive oil (300 and 3,000 mg α-tocopherol/kg lipid)	4 °C, air, active MAP (20 % O_2, 40 % O_2, 60 % O_2, 80 % O_2)	O'Grady et al. (2000)
Beef patties	Ascorbic acid (500 ppm), taurine (50 mM), carnosine (50 mM), rosemary powder (1,000 ppm), and their combinations	2 °C, active MAP (70 % O_2, 20 % CO_2, 10 % N_2)	Sanchez-Escalante et al. (2001)
Minced meat	Oregano essential oil (0–1 % v/w)	5 °C, air, active MAP (30 % O_2, 40 % CO_2, 30 % N_2, 100 % CO_2)	Skandamis and Nychas (2001)
Beef patties	Ascorbic acid, oregano extract, lycopene rich tomato pulp, and their mixtures	2 °C, active MAP (70 % O_2, 20 % CO_2, 10 % N_2)	Sanchez-Escalante et al. (2003)
Minced beef patties	Tea catechins (0–1,000 ppm)	4 °C, air, active MAP (80 % O_2, 20 % CO_2, 0 % N_2)	Tang et al. (2006)
Beef patties	Rosemary extract (0.05 %) ascorbate/citrate (0.05 %)	4 °C, air, active MAP (100 % N_2, 80 % O_2, 20 % N_2)	Lund et al. (2007)
Nonconventional poultry patties	Thymol (0–300 ppm) Carvacrol (0–300 ppm)	0–18 °C, air, active MAP (40 % CO_2, 30 %O_2, 30 % N_2)	Mastromatteo et al. (2009)
Fresh meat patties	Thymol (0–750 ppm)	4 °C, air and active MAP (40 % O_2, 15 % CO_2, 45 % N_2)	Del Nobile et al. (2009a)

are still needed for successful MAP systems intended for meat applications. The key to successful packaging is the selection of materials and designs that best balance the competing needs of product characteristics, marketing considerations including distribution and consumer needs, environmental and waste management issues, and cost (Marsh and Bugusu 2007).

Scientific knowledge on the effectiveness of combinations of various hurdles, especially the popular MAP technology with novel antimicrobial substances, further promotes their applications to fresh and processed meat (Mastromatteo et al. 2010). The combination of MAP and active compounds of different origins seems to be the most diffuse strategy for preserving fresh meat in the form of hamburgers, meatballs, and patties (Table 9.1).

Del Nobile et al. (2009a) evaluated the effectiveness of thymol, combined with other extrinsic factors, storage temperatures, and MAP, to extend the microbiological SL of packed fresh meat patties. The product was supplemented

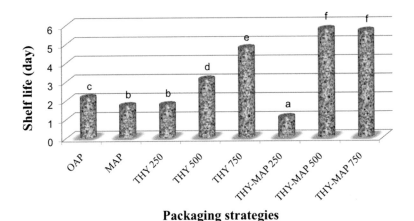

Packaging strategies

Fig. 9.1 Shelf life values (day) of minced beef patties packaged under different conditions. Statistically significant differences ($p < 0.05$) between samples of each group are identified by different letters. OAP = sample packaged under ordinary atmosphere; MAP = sample packaged under modified atmosphere (40:15:45 O_2:CO_2:N_2); THY250 = sample with 250 mg/kg thymol added; THY500 = sample with 500 mg/kg thymol added; THY750 = sample with 750 mg/kg thymol added; THY-MAP 250 = sample with 250 mg/kg thymol added and packaged under MAP (40:15:45 O_2:CO_2:N_2); THY-MAP 500 = sample with 500 mg/kg thymol added and packaged under MAP (40:15:45 O_2:CO_2:N_2); THY-MAP 750 = sample with 750 mg/kg thymol added and packaged under MAP (40:15:45 O_2:CO_2:N_2)

with thymol at levels of 250, 500, and 750 mg/kg ground beef. The treated and untreated meat samples were packaged in high-barrier films under ordinary and MAP (40:15:45 O_2:CO_2:N_2) conditions for 16 days. Results showed that thymol, working alone, was effective on coliforms and Enterobacteriaceae, whereas it did not seem to inhibit very much the growth of other microbial populations. Conversely, high thymol concentration, combined with MAP conditions, highly affected product quality, thus allowing a prolongation of the SL from 2 to 5 days compared to the control samples (Fig. 9.1). Mastromatteo et al. (2009) also studied the combined effect of thymol (0–300 ppm), carvacrol (0–300 ppm), and temperature on the microbiological quality of nonconventional poultry patties packaged under ordinary and modified atmospheres (30:40:30 O_2:CO_2:N_2). The results showed that the final total viable count for patties packaged in air decreased with the increase in carvacrol up to 150 ppm, and for higher carvacrol concentrations an increase in the microbial population was observed. In the case of samples treated with thymol the reduction in the activity at increasing concentrations of the active compound was less evident. The combination of MAP and active agents had better effects on microbial counts, thus also limiting the off-odors associated with spoilage.

The novel concept of packaging is represented by active packaging as a system that can extend SL or improve safety or sensory properties while maintaining the quality of food (Lee et al. 2008). The concept, which is successfully utilized in the

USA and Japan, has seen only limited development in Europe. This could be due to legal restrictions, to a lack of knowledge about both the acceptability of these systems to consumers and their effectiveness in packaging, or to their economic and environmental impacts (Coma 2008). The applications of active packaging to fresh meat are extensively reviewed in various works, with a special emphasis placed on antimicrobial packaging systems (Aymerich et al. 2008; Coma 2008; McMillin 2008; Quintavalla and Vicini 2002). The principal active packaging systems include oxygen scavengers, control of CO_2 or ethylene concentration and generation, ethanol and antioxidant releasers, moisture absorption or desiccants, and antimicrobial systems to control several microorganisms. These latter active systems can be divided into five main categories: (1) systems based on the direct incorporation of antimicrobial substances into the packaging film; (2) systems based on the incorporation of antimicrobial substances into a sachet in the package; (3) systems based on coating of the packaging with a material that acts as a carrier for the additive; (4) antimicrobial systems based on macromolecules with film-forming properties; (5) systems based on bioactive edible coatings directly applied to the food, as also reported in Chaps. 4 and 5 of Part II of this book. Organic acids, bacteriocins, enzymes, spices, and essential oils and other compounds as triclosan or tocopherol are the most diffuse active agents adopted for packaging beef carcasses, cooked hams, fresh poultry breasts, and cured meats (Coma 2008). The specific application of antimicrobial packaging to minced meat are not as abundant (Gennadios et al. 1997; Mauriello et al. 2004), probably due to the fact that direct incorporation of preservatives in minced meat, hamburger or patties, results in a more immediate reduction in the bacterial population compared to the effects exerted by release systems. For packaging with bioactive compounds dispersed in a film or coated on a polymeric surface and with a nonvolatile biocide, contact between the food and the package is obviously necessary, and a migration process from the packaging materials is expected. Moreover, processed meat generally has an internal contamination that cannot be controlled by packaging systems that act on product surfaces. Therefore, potential food applications of active packaging include, in particular, vacuum-packed products and meat with surface contamination (Millette et al. 2007; Samelis et al. 2005; Yingyuad et al. 2006).

Thermal alternative technologies such as high-frequency heating, including microwave and radiofrequency, ohmic heating or steam pasteurization applied to the product after final packaging could bring new possibilities to the pasteurization of meat products, even if these systems are more suitable for ready-to-eat meal. Microbial inactivation of meat products by high-frequency heating has been widely demonstrated in the literature (Orsat and Raghavan 2005; Jeong et al. 2006; Picouet et al. 2007). Moreover, the possibility of offering continuous systems is seen as an advantage in food processing, although there are still nonuniformity problems to be solved (Aymerich et al. 2008). Similar limitations are also present in steam pasteurization. The efficiency of this treatment for meat products and meat pieces has been presented in a series of articles; however, a prolonged treatment of longer than 10 s gives a cooked appearance to the sample, thus limiting the application

(McCann et al. 2006). As regards ohmic heat technology, it has also proved to be a successful technology to process liquids, but its use in solid meat products has not yet found industrial application (Piette et al. 2004).

9.3 Fish Burgers

In recent years demand for fish products has been on the rise, especially for their high nutritional value and peculiarities that make fish products different from other foods (Paleologos et al. 2004). The numerous studies conducted to date have pointed out that fish consumption is linked to healthy–nutritional motivations for home consumption and to hedonistic motivations outside the home. In response to the increased demand for fish products, increased fishing has provoked a rise in the wasting of fish resources due to unsuitable fishing practices. In fact, fishing nets, trains, traps, and fishing lines cause bycatch, that is, the mindless fishing of products that, being outside the range of commercially desirable sizes, are discarded because they are unsuitable for direct consumption (Cabral et al. 2003). This problem represents an important menace for marine environments because casual fishing produces a bycatch amounting to one million tons a year. The resulting damage has many aspects: the overharvesting of the more common species, the lack of valorization of those species whose nutritional and sensory value has been forgotten, and an increase in economic problems in Italian fishing.

The lack of consumption of particular fishes depends on the poor knowledge of the products, their nutritional characteristics, the useful storage conditions to prolong the SL, the optimal conditions to clean and cook the products, their sensory characteristics, and the acceptability and consumer taste (Del Nobile et al. 2009b; Wakland et al. 2005).

Nutritionists have always pointed out the importance of fish in the human diet, and in particular blue fish, due to the presence of protein of high biological value and a fatty acids rich in polyunsatured fatty acids (omega-3) (Simopoulos 1989). The consumption of at least two to three meals of fish a week, in a balanced diet, may play an important role in the prevention of numerous diseases. The Italian Society of Human Nutrition indicates a minimum level of polyunsaturated fatty acid intake, and in particular of omega-3 fatty acids from fish, fish oils, and crustaceans, equal to 0.5 % of total daily calories (Singer et al. 1986; Soriguer et al. 1997; Steffens 1997).

For a long time, researchers have been interested in problems related to the microbial contamination of fish and the consequent quality loss (Dalgaard 1995; Gram and Huss 1996; Paleologos et al. 2004; Gonzalez-Rodriguez et al. 2001). Several articles have appeared on the biopreservation of fresh fish, especially on the use of lactic acid bacteria and their metabolites (Al-Dagal and Bazaraa 1999; Altieri et al. 2005), and on the use of MAP to prolong SL (Amanatidou et al. 2000; Corbo et al. 2005; Goulas and Kontominas 2007; Poli et al. 2006; Sivertsvik et al. 2002; Torrieri et al. 2006).

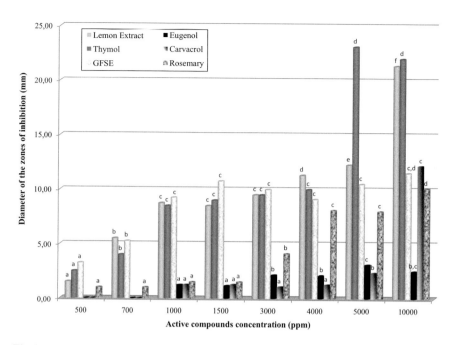

Fig. 9.2 Inhibitory effect of natural compounds tested against *Shewanella putrefaciens* by in vitro test. Statistically significant differences ($p < 0.05$) between samples of each group are identified by different letters

A review of the literature shows that, to date, few attempts have been made to valorize fish characterized by a low economic value (Gildberg 2001, 2004; Lòpez-Caballero et al. 2005). Taking into account consumer opinion on fresh fish as a time-consuming product (Trondsen et al. 2003), the development of ready-to-cook meals based on fish could represent a strategic solution. Therefore, to boost fish consumption, valorize low-commercial-value fish, and increase knowledge about fish products characterized by a high *convenience*, a group of researchers from Southern Italy investigated the possibility of producing and preserving new fish burgers. The experimental plans carried out are briefly described in what follows. Corbo et al. (2009b) carried out an in vitro test on the effects of carvacrol, eugenol, thymol, green tea extract, rosemary extract, grapefruit seed extract, and lemon extract, at various concentrations ranging from 500 to 10,000 ppm, on the main fish spoilage microorganisms (*Shewanella putrefaciens* and *Photobacterium phosphoreum*). As can be seen in Fig. 9.2, in a comparison of the effectiveness of active compounds against *S. putrefaciens*, thymol showed the greatest inhibition activity. For its efficacy against both microbial targets, thymol was used as a natural preservative in fish burgers composed of cod (*Gadus morhua*). The samples were packaged in barrier bags under ordinary and MAP conditions (100 % of CO_2) and stored at 4 °C for 9 days. Results obtained by the authors suggested that the use of thymol combined with MAP was effective at improving the microbiological stability of cod burgers. Compared with several other mild preservation procedures, like low-dose irradiation, addition of protective cultures, or high-pressure treatment, the proposed preservation technique could be an inexpensive and uncomplicated method for extending SL. Ouattara et al. (2001) also found that SL of shrimp was extended using a combined treatment of gamma irradiation, thymol oil, and trans-cinnamaldehyde. Mejholm and Dalgaard (2002) observed that oregano oil reduced the growth of *P. phosphoreum* and extended the SL of cod fillets kept under MAP. Harpaz et al. (2003) observed that treatment with oregano (0.05 %) or thyme increased the SL of Asian sea bass fish. Mahmoud et al. (2004, 2006) observed that dipping carp fillets into a solution containing both carvacrol and thymol, with and without a combined treatment with electrolyzed NaCl, led to an extension of SL. In a subsequent experimental trial, Corbo et al. (2009c) tested the effects of thymol, lemon extract, and citrus extract, at 20, 40, and 80 ppm, on the main spoilage microorganisms inoculated in fish burgers, based on sea bream (*Sparus auratus*), stored at 5 °C. By means of an appropriate data elaboration, the authors assessed product acceptability as the lowest value between the microbial and the sensory quality. All investigated compounds were found to be effective in slowing down microbial growth. In particular, citrus extract and thymol were the most effective against the specific spoilage bacteria of marine temperate-water fish. Since microbial proliferation appeared to be the factor limiting fresh fish burger quality, a further effort to slow down microbial spoilage was carried out by combining the three compounds through a central composite design (CCD) (Table 9.2) (Corbo et al. 2008). To this end, the cell growth of the main fish spoilage microorganisms (*Pseudomonas fluorescens*, *Photobacterium phosphoreum*, and *Shewanella putrefaciens*), inoculated in fish burgers, and the growth of mesophilic and psychrotrophic bacteria were monitored. The most striking feature of the results is not only the increase in the MAL value for burgers mixed with the antimicrobial compounds, compared to the control sample (Table 9.2), but also the possibility of determining the optimal antimicrobial compound composition. Higher microbiological quality, with no unpleasant flavor, was assured by combining 110 mg L^{-1} thymol, 100 mg L^{-1} citrus extract, and 120 mg L^{-1} lemon extract.

To produce nutritionally balanced fish burgers, a chemical investigation on chub mackerel and hake was conducted with the aim of mixing the two fish species in a new fish burger (Di Monaco et al. 2009). Focus group interviews and consumer tests were also performed to evaluate consumer opinion about these new fresh products and best acceptability of various proportions of mackerel and hake. The results of chemical analyses proved that the desired nutritional properties of fish were assured by high amounts of chub mackerel, whereas a low level of hake had to be used to obtain firmer and wetter burgers. In addition, the consumer test showed that the most successful sample was that with 70 % chub mackerel and 30 % hake.

Bearing in mind that MAP is the most widespread preservation technology for fresh fish (Boskou and Debevere 2000; Corbo et al. 2005; Devlieghere et al. 2004; Poli et al. 2006; Torrieri et al. 2006; Sivertsvik et al. 2002; Tang et al. 2001) and in view of the numerous scientific publications dealing with the combined effects of

Table 9.2 Microbiological acceptability limit (MAL) of fish burgers, prepared by mixing sea bream with various concentrations of active compounds. MAL values were calculated by fitting a reparameterized version of Gompertz equation to experimental data relative to *Shewanewlla putrefaciens*. MAL value corresponds to time to attain *S. putrefaciens* count value of 10^6 CFU g^{-1}. Data are presented $\pm$ standard error

	Thymol (ppm w/v)	Citrus extract (ppm v/v)	Lemon extract (ppm v/v)	MAL (day)
Control	0	0	0	2.0725 ± 0.0418
Run 1	150	150	150	3.9805 ± 0.0518
Run 2	50	150	150	3.8041 ± 0.0484
Run 3	150	150	50	3.2057 ± 0.0565
Run 4	50	150	50	3.0647 ± 0.0126
Run 5	150	50	150	3.1429 ± 0.0175
Run 6	50	50	150	3.1186 ± 0.0207
Run 7	150	50	50	2.9601 ± 0.0748
Run 8	50	50	50	3.0429 ± 0.0237
Run 9	100	100	100	3.4177 ± 0.0972
Run 10	100	200	100	2.8872 ± 0.0625
Run 11	100	0	100	2.7289 ± 0.0200
Run 12	100	100	200	2.7102 ± 0.0289
Run 13	100	100	0	2.3767 ± 0.0545
Run 14	200	100	100	3.4652 ± 0.0309
Run 15	0	100	100	2.6120 ± 0.0484
Run 16	100	100	100	3.4177 ± 0.0972
Run 17	100	100	100	3.4177 ± 0.0972

MAP and active compounds (Corbo et al. 2009a; Masniyom et al. 2006; Mastromatteo et al. 2010), Del Nobile et al. (2009b) assessed the SL prolongation of fish burgers made up of 70 % chub mackerel and 30 % hake, produced using active compounds and packaged under MAP conditions. In the study, the three previously tested essential oils (thymol and two citrus extracts) were combined with three different gas mix compositions (30:40:30 $O_2:CO_2:N_2$, 50:50 $O_2:CO_2$ and 5:95 $O_2:CO_2$). Burger fish were packaged and stored at 4 °C for 28 days. During storage, the nutritional, microbiological, and sensory quality, together with the development of biogenic amines, was assessed. The microbiological results highlighted a synergistic effect between active compounds and MAP, especially using the gas headspace with high amounts of CO_2. From a sensory point of view, an unpleasant odor was perceived after about 3 weeks. Figure 9.3 reports the overall quality of samples packaged in the various MAP conditions. As can be inferred from the figure, the maximum sensory acceptability value accounts for approximately 25 days in the product under 100 % CO_2. Under these last-named packaging conditions the microbiological acceptability was approximately 28 days, and hence product SL was limited by sensory properties to approximately 25 days. At any rate, a fish burger with a SL of more than 3 weeks represents an interesting goal for a fresh-based product, worthy of industry attention, and suitable to promote fish consumption beyond local borders.

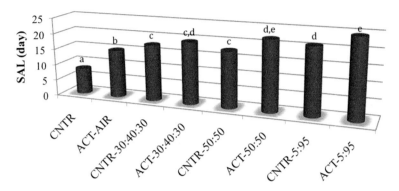

Fig. 9.3 Sensorial acceptability limit (SAL) (day) of fish burgers packaged under different conditions. SAL values were calculated by fitting a reparameterized version of Gompertz equation to experimental data. SAL value corresponds to the time to attain a score of 2, using a scale ranging from 0 to 5 (where 0 = very poor and 5 = excellent). Statistically significant differences ($p < 0.05$) between samples are identified by different letters. CNTR = untreated fish burger sample packaged under air; ACT-AIR = fish burger with active compounds packaged under air; CNTR-30:40:30 = fish burger without active compounds packaged under MAP (30:45:30 $O_2:CO_2:N_2$); ACT-30:40:30 = fish burger with active compounds packaged under MAP (30:45:30 $O_2:CO_2:N_2$); CNTR-50:50 = fish burger without active compounds packaged under MAP (50:55:0 $O_2:CO_2:N_2$); ACT-50:50 = fish burger with active compounds packaged under MAP (50:50:0 $O_2:CO_2:N_2$); CNTR-5:95 = fish burger without active compounds packaged under MAP (5:95:0 $O_2:CO_2:N_2$); ACT-5:95 = fish burger with active compounds packaged under MAP (5:95:0 $O_2:CO_2:N_2$)

References

Ahn DU, Nam KC (2004) Effects of ascorbic acid and antioxidants on colour, lipid oxidation and volatiles of irradiated ground beef. Radiat Phys Chem 71:149–154

Al-Dagal MM, Bazaraa WA (1999) Extension of shelf life of whole and peeled shrimp with organic acid salts and bifidobacteria. J Food Prot 62:51–56

Altieri C, Speranza B, Del Nobile MA, Sinigaglia M (2005) Suitability of bifidobacteria and thymol as biopreservatives in extending the shelf life of fresh packed plaice fillets. J Appl Microbiol 99:1294–1302

Amanatidou A, Schluter O, Lemkau K, Gorris LGM, Smid EJ, Knorr D (2000) Effect of combined application of high pressure treatment and modified atmospheres on the shelf life of fresh Atlantic salmon. Innov Food Sci Emerg Technol 1:87–98

Aymerich T, Picouet PA, Monfort JM (2008) Decontamination technologies for meat products. Meat Sci 78:114–129

Belcher JN (2006) Industrial packaging developments for the global meat market. Meat Sci 74:143–148

Bernbom N, Licht TR, Brogren CH, Jelle B, Johansen AH, Badiola I, Vogensen FK, Norrung B (2006) Effects of Lactobacillus lactis on composition of intestinal microbiota: role of nisin. Applied Environ Microbiol 72:239–244

Boskou G, Debevere J (2000) Shelf life extension of cod fillets with an acetate buffer spray prior to packaging under modified atmosphere. Food Addit Contam 17:17–25

Cabral H, Duque J, Costa MJ (2003) Discards of the beach seine fishery in the central coast of Portugal. Fish Res 63:63–71

Cannarsi M, Baiano A, Sinigaglia M, Ferrara I., Bacalo R, Del Nobile MA (2008) Use of nisin, lysozyme and EDTA for inhibiting microbial growth in chilled buffalo MEAT. Int J Food Sci Technol 43:573–578

Coma V (2008) Bioactive packaging technologies for extended shelf life of meat-based products. Meat Sci 78:90–103

Corbo MR, Altieri C, Bevilacqua A, Campaniello D, D'Amato D, Sinigaglia M (2005) Estimating packaging atmosphere-temperature effects on the shelf life of cod fillets. Eur Food Res Technol 220:509–513

Corbo MR, Speranza B, Filippone A, Granatiero S, Conte A, Sinigaglia M, Del Nobile MA (2008) Study on the synergic effect of natural compounds on the microbial quality decay of packed fish hamburger. Int J Food Microbiol 127:261–267

Corbo MR, Bevilacqua A, Campaniello D, D'Amato D, Speranza B, Sinigaglia M (2009a) Prolonging microbial shelf life of foods through the use of natural compounds and non-thermal approaches – a review. Int J Food Sci Technol 44:223–241

Corbo MR, Di Giulio S, Conte A, Speranza B, Sinigaglia M, Del Nobile MA (2009b) Thymol and modified atmosphere packaging to control microbiological spoilage in packed fresh cod hamburgers. Int J Food Sci Technol 44:1553–1560

Corbo MR, Speranza B, Filippone A, Conte A, Sinigaglia M, Del Nobile MA (2009c) Natural compounds to preserve fresh fish burgers. Int J Food Sci Technol 44:2021–2027

Dalgaard P (1995) Qualitative and quantitative characterization of spoilage bacteria from packed fish. Int J Food Microbiol 26:319–333

Darmadji P, Izumimoto M (1994) Effect of chitosan in meat preservation. Meat Sci 38:243–254

Del Nobile MA, Conte A, Cannarsi M, Sinigaglia M (2009a) Strategies for prolonging the shelf life of minced beef patties. J Food Saf 29:14–25

Del Nobile MA, Corbo MR, Speranza B, Sinigaglia M, Conte A, Caroprese M (2009b) Combined effect of MAP and active compounds on fresh blue fish burger. Int J Food Microbiol 135:281–287

Devlieghere F, Vermeiren L, Debevere J (2004) New preservation technologies: possibilities and limitations. Int Dairy J 14:273–285

Di Monaco R, Cavella S, Masi P, Sevi A, Caroprese M, Marzano A, Conte A, Del Nobile MA (2009) Fresh burgers based on blue fish: nutritional and sensory optimization. Int J Food Sci Technol 44:1634–1641

Dzudie T, Kouebou CP, Essia-Ngang JJ, Mbofung CMF (2004) Lipid sources and essential oils effects on quality and stability of beef patties. J Food Eng 65:67–72

Ellis M, Cooksey K, Dawson P, Han I, Vergano P (2006) Quality of fresh chicken breasts using a combination of modified atmosphere packaging and chlorine dioxide sachets. J Food Prot 69:1991–1996

Fernández-Ginés JM, Fernández-López J, Sayas-Barberá E, Pérez-Alvarez JA (2005) Meat products as functional foods: a review. J Food Sci 70:37–43

Fernández-López J, Zhi N, Aleson-Carbonell L, Perez-Alvarez JA, Kuri V (2005) Antioxidant and antibacterial activities of natural extracts: Application in beef meatballs. Meat Sci 69:371–380

Gennadios A, Hanna MA, Kurth LB (1997) Application of edible coatings on meats, poultry and seafoods: a review. Lebensmittel-Wissenschaft und Technology 30:337–350

Gildberg A (2001) Utilisation of male Arctic capelin and Atlantic cod intestines for fish sauce production-evaluation of fermentation conditions. Bioresour Technol 76:119–123

Gildberg A (2004) Digestive enzyme activities in starved pre-slaughter farmed and wild-captured, Atlantic cod (Gadus morhua). Aquaculture 238:343–353

Gola S, Mutti P, Manganelli E, Squarcina N, Rovere P (2000) Behaviour of *E. coli* O157:H7 strains in model system and in raw meat by HPP: microbial and technological aspects. High Press Res 19:481–487

González-Fandos E, Villarino-Rodríguez A, García-Linares MC, García-Arias MT, García-Fernández MC (2005) Microbiological safety and sensory characteristics of salmon slices processed by the sous vide method. Food Control 16:77–85

Gonzalez-Rodriguez MN, Sanz JJ, Santos JA, Otero A, Garcia-Lopez ML (2001) Bacteriological quality of aquaculture fresh-water fish portions in prepackaged trays stored at 3°C. J Food Prot 64:1399–1404

Goulas AE, Kontominas MG (2007) Combined effect of light salting, modified atmosphere packaging and oregano essential oil on the shelf-life of sea bream (Sparus aurata): biochemical and sensory attributes. Food Chem 100:287–296

Gram L, Dalgaard P (2002) Fish spoilage bacteria-problems and solutions. Curr Opin Biotechnol 13:262–266

Gram L, Huss HH (1996) Microbiological spoilage of fish and fish products. Int J Food Microbiol 33:121–137

Harpaz S, Glatman L, Drabkin V, Gelman A (2003) Effects of herbal essential oils used to extend the shelf life of freshwater-reared Asian sea bass fish (Lates calcarifer). J Food Prot 66:410–417

Hugas M (1998) Bacteriocinogenic lactic acid bacteria for the biopreservation of meat and meat products. Meat Sci 49:139–150

Hugas M, Pagés F, Garriga M, Monfort JM (1998) Application of the bacteriocinogenic Lactobacillus sakei CTC494 to prevent growth of Listeria in fresh and cooked meat products packed with different atmospheres. Food Microbiol 15:639–650

Hugas M, Garriga M, Monfort JM (2002) New mild technologies in meat processing: high pressure as a model technology. Meat Sci 62:359–371

Ingram M, Dainty RH (1971) Changes caused by microbes in spoilage of meats. J Appl Bacteriol 34:21–39

Jacobsen M, Bertelsen G (2000) Color stability and lipid oxidation of fresh beef. Development of a response surface model for predicting the effects of temperature, storage time and modified atmosphere composition. Meat Sci 54:49–57

Jacobsen M, Bertelsen G (2002) The use of CO_2 in packaging of fresh red meats and its effect on chemical quality changes in the meat: a review. J Muscle Foods 13:143–168

Jacobsen T, Budde BB, Koch AG (2003) Application of Leuconostoc carnosum for biopreservation of cooked meat products. J Appl Microbiol 95:242–249

Jeong JY, Lee ES, Choi JH, Lee JY, Kim JM, Min SG et al (2006) Variability in temperature distribution and coking properties of ground pork patties containing different fat level and with/without salt cooked in microwave energy. Meat Sci 75:415–422

Jo C, Son JH, Son CB, Byun MW (2003) Functional properties of raw and cooked pork patties with added irradiated, freeze-dried green tea leaf extract powder during storage at 4°C. Meat Sci 64:17–33

Kerry JP, O'Grady MN, Hogan SA (2006) Past, current and potential utilization of active and intelligent packaging systems for meat and muscle-based products: a review. Meat Sci 74:113–130

Lee S, Hernández P, Djordjevic D, Faraji H, Hollender R, Faustman C et al (2006a) Effect of antioxidants and cooking on stability of n-3 fatty acids in fortified meat products. J Food Sci 71:233–238

Lee S, Faustman C, Djordjevic D, Faraji H, Decker EA (2006b) Effect of antioxidants on stabilization of meat products fortified with n-3 fatty acids. Meat Sci 72:18–24

Lee DS, Yam KL, Piergiovanni L (2008) Food packaging science and technology. CRC Press, Boca Raton

Leistner L (2000) Basic aspects of food preservation by hurdle technology. Int J Food Microbiol 55:181–186

Lòpez-Caballero ME, Gòmez-Guillén MC, Pérez-Mateos M, Montero P (2005) A chitosan–gelatin blend as a coating for fish patties. Food Hydrocolloid 19:303–311

Lund MN, Hviid MS, Skibsted LH (2007) The combined effect of antioxidants and modified atmosphere packaging on protein and lipid oxidation in beef patties during chill storage. Meat Sci 76:226–233

Mahmoud BSM, Yamazaki K, Miyashita K, Shin IS, Chong D, Suzuki T (2004) Bacterial microflora of carp (Cyprinus carpio) and its shelf life extension by essential oil compounds. Food Microbiol 21:657–666

Mahmoud BSM, Yamazaki K, Miyashita K, Shin IS, Chong D, Suzuki T (2006) A new technology for fish preservation by combined treatment with electrolyzed NaCl solutions and essential oil compounds. Food Chem 99:656–662

Marsh K, Bugusu B (2007) Food packaging – roles, materials, and environmental issues. Food Sci 72:39–55

Masniyom P, Benjakul S, Visessanguan W (2006) Synergistic antimicrobial effect of pyrophosphate on Listeria monocytogenes and Escherichia coli O157 in modified atmosphere packaged and refrigerated seabass slices. LWT Food Sci Technol 39:302–307

Mastromatteo M, Lucera A, Sinigaglia M, Corbo MR (2009) Combined effects of thymol, carvacrol and temperature on the quality of non conventional poultry patties. Meat Sci 83:246–254

Mastromatteo M, Conte A, Del Nobile MA (2010) Combined use of modified atmosphere packaging and natural compounds for food preservation. Food Eng Rev 2:28–38

Mauriello G, Ercolini D, La Storia A, Casaburi A, Villani F (2004) Development of polythene films for food packaging activated with an antilisterial bacteriocin from Lactobacillus curvatus 32Y. J Appl Microbiol 97:314–322

McCann MS, Sheridan JJ, McDowell DA, Blair IS (2006) Effect of steam pasteurization on Salmonella Typhimurium DT104 and Escherichia coli O157: H7 surface inoculated onto beef, pork and chicken. J Food Eng 76:32–40

McMillin KV (2008) Where is MAP going? A review and future potential of modified atmosphere packaging for meat. Meat Sci 80:43–65

Mejlholm O, Dalgaard P (2002) Antimicrobial effect of essential oils on the seafood spoilage microorganism Photobacterium phosphoreum in liquid media and fish products. Lett Appl Microbiol 34:27–31

Millette M, Le Tien C, Smoragiewicz W, Lacroix M (2007) Inhibition of Staphylococcus aureus on beef by nisin-containing modified alginate films and beads. Food Control 18:878–884

Mitsumoto M, O'Grady MN, Kerry JP, Buckley DJ (2005) Addition of tea catechins and vitamin C on sensory evaluation, colour and lipid stability during chilled storage in cooked or raw beef and chicken patties. Meat Sci 69:773–779

Nissen LR, Byrne DV, Bertelsen G, Skibsted LH (2004) The antioxidative activity of plant extracts in cooked pork patties as evaluated by descriptive sensory profiling and chemical analysis. Meat Sci 68:485–495

O'Grady MN, Monahan FJ, Burke RM, Allen P (2000) The effect of oxygen level and exogenous α-tocopherol on the oxidative stability of minced beef in modified atmosphere packs. Meat Sci 55:39–45

Orsat V, Raghavan V (2005) Microwave technology for food processing: an overview in the microwave processing of foods. Wood-head, Cambridge, pp 105–118

Ouattara B, Sabato SF, Lacroix M (2001) Combined effect of antimicrobial coating and gamma irradiation on shelf life extension of pre-cooked shrimp (Penaeus spp.). Int J Food Microbiol 68:1–9

Paleologos EK, Savvaidis IN, Kontominas MG (2004) Biogenic amines formation and its relation to microbiological and sensory attributes in ice-stored whole, gutted and filleted Mediterranean Sea bass (Dicentrarchus labrax). Food Microbiol 21:549–557

Pawar DD, Malik SVS, Bhilegaonkar KN, Barbuddhe SB (2000) Effect of nisin and its combination with sodium chloride on the survival of Listeria monocytogenes added to raw buffalo meat mince. Meat Sci 56:215–219

Picouet PA, Fernández A, Serra X, Suñol JJ, Arnau A (2007) Microwave heating of cooked pork patties as a function of fat content. J Food Sci 72:57–63

Piette G, Buteau ME, de Halleux D, Chiu L, Raymond Y, Ramaswany HS et al (2004) Ohmic cooking of processed meat and its effects on product quality. J Food Sci 69:71–78

Poli MB, Messini A, Parisi G, Scappini F, Figiani V (2006) Sensory, physical, chemical and microbiological changes in European sea bass (Dicentrarchus labrax) fillets packed under modified atmosphere/air or prepared from whole fish stored in ice. Int J Food Sci Technol 41:444–454

Quintavalla S, Vicini L (2002) Antimicrobial food packaging in meat industry. Meat Sci 62:373–380

Racanicci AMC, Danielsen B, Menten JFM, Regitano-d'Arce MAB, Skibsted LH (2004) Antioxidant effect of dittany (Origanum dictamnus) in pre-cooked chicken meat balls during chillstorage in comparison to rosemary (Rosmarinus officinalis). Eur Food Res Technol 218:521–524

Sagoo S, Board R, Roller S (2002) Chitosan inhibits growth of spoilage micro- organisms in chilled pork products. Food Microbiol 19:175–182

Samelis J, Bedie GK, Sofos JN, Belk KE, Scanga JA, Smith GC (2005) Combinations of nisin with organic acids or salts to control Listeria monocytogenes on sliced pork bologna stored at 4°C in vacuum packages. Lebensm Wiss Technol 38:21–28

Sanchez-Escalante A, Djenane D, Torrescano G, Beltran JA, Roncales P (2001) The effects of ascorbic acid, taurine, carnosine and rosemary powder on colour and lipid stability of beef patties packaged in modified atmosphere. Meat Sci 58:421–429

Sanchez-Escalante A, Torrescano G, Djenane D, Beltran JA, Roncales P (2003) Combined effect of modified atmosphere packaging and addition of lycopene rich tomato pulp, oregano and ascorbic acid and their mixtures on the stability of beef patties. Food Sci Technol Int 9:77–84

Schillinger U, Kaya M, Lücke FK (1991) Behaviour of Listeria monocytogenes in meat and its control by bacteriocin-producing strain of Lactobacillus sake. J Appl Bacteriol 70:473–478

Simopoulos AP (1989) Summary of NATO advanced research workshop on dietary ώ3 and ώ6 fatty acids biological effects and nutritional essentiality. J Nutr 199:512–528

Singer P, Berger I, Luck K, Taube C, Naumann E, Godick W (1986) Long-term effect of mackerel diet on blood pressure, serum lipids and thromboxane formation in patients with mild essential hypertension. Atherosclerosis 62:259–265

Sivertsvik M, Jeksrud WK, Rosnes JT (2002) A review of modified atmosphere packaging of fish and fishery products – significance of microbial growth, activities and safety. Int J Food Sci Technol 37:107–127

Skandamis PN, Nychas G-JE (2001) Effect of oregano essential oil on microbiological and physicochemical attributes of minced meat stored in air and modified atmospheres. J Appl Microbiol 91:1011–1022

Skandamis P, Tsigarida E, Nychas G-JE (2000) Ecophysiological attributes of Salmonella typhimurium in liquid culture and within gelatin gel with or without the addition of oregano essential oil. World J Microbiol Biotechnol 16:31–35

Sommers C, Boyd G (2006) Variations in the radiation sensitivity of foodborne pathogens associated with complex ready-to-eat food products. Radiat Phys Chem 75:773–778

Soriguer F, Serna S, Valverde E, Hernando J, Martín-Reyes A, Soriguer M, Pareja A, Tinahones F, Esteva I (1997) Lipid, protein, and calorie content of different Atlantic and Mediterranean fish, shellfish, and molluscs commonly eaten in the south of Spain. Eur J Epidemiol 13:451–463

Stahl NZ (2007) New packaging formats drive new products. Meat Process 46:40–42, 44

Steffens W (1997) Effects of variation in essential fatty acids in fish feeds on nutritive value of freshwater fish for humans. Aquaculture 151:97–119

Tang SZ, Kerry JP, Sheehan D, Buckley DJ, Morrissey PA (2001) Antioxidative effect of added tea catechins on susceptibility to oxidative stability in cooked red meat, poultry and fish patties. Food Res Int 34:651–657

Tang SZ, Ou SY, Huang SX, Li W, Kerry JP, Buckley DJ (2006) Effects of added tea catechins on colour stability and lipid oxidation in minced beef patties held under aerobic and modified atmospheric packaging conditions. J Food Eng 77:248–253

Torrieri E, Cavella S, Villani F, Masi P (2006) Influence of modified atmosphere packaging on the chilled shelf life of gutted farmed bass (Dicentrarchus labrax). J Food Eng 77:1078–1086

Trondsen T, Scholderer J, Lund E, Eggen AE (2003) Perceived barriers to consumption of fish among Norwegian women. Appetite 41:301–314

USDA (1995) U.S. Department of Agriculture. Sample collection. Ms K. Sloan and Ms F. Pathogen reduction; hazard analysis and critical control point (HACCP) systems; proposed Dusseault provided skilled technical assist- rule. Fed Regist 60:6774–6889

Vuorela S, Salminen H, Makela M, Kivikari R, Karonen M, Heinonen M (2005) Effect of plant phenolics on protein and lipid oxidation in cooked pork meat patties. J Agric Food Chem 53:8492–8497

Wakland HM, Stoknes IS, Remme JF, Kjerstad M, Synnes M (2005) Proximate composition, fatty acid and lipid class composition of the muscle from deep-sea teleosts and elasmobranches. Comp Biochem Physiol 140:437–443

Xiong YL (2000) Protein oxidation and implications for muscle food quality. In: Decker E, Faustman C (eds) Antioxidants in muscle foods. Wiley, Chichester, pp 85–111

Yin MC, Faustman C (1993) Influence of temperature, pH and phospholipid composition upon the stability of myoglobin and phospholipid: a liposome model. J Agric Food Chem 41:853–857

Yingyuad S, Ruamsin S, Reekprkhon D, Douglas S, Pongamphai S, Siripatrawan U (2006) Effect of chitosan coating and vacuum packaging on the quality of refrigerated grilled pork. Packag Technol Sci 19:149–157

Future Perspective

During the early twentieth century, substantial improvements were made to both rigid and flexible packaging materials, thereby increasing significantly the options available for maintaining the quality and improving the shelf life of foods. Efforts to improve the performances of packaging solutions and control fresh foods can be directed toward many areas. Tamper evidence and closable features are seen as important factors in improving the performance of packaging. One of the requirements for food packaging was that it should play a passive role, remaining inert and not interacting with the food it contains. However, the development of active packaging now makes it acceptable for the packaging to have a more interactive role in extending the shelf life of foods. This means that active systems will be used more widely to enhance the shelf life of products. Even though much progress has been made, further research still needs to be conducted on how various active packaging solutions affect product characteristics. Participation and collaboration of research institutions, industry, and government regulatory agencies will be key in the success of active packaging technologies for food applications. More work in this regard will expand the applicability and further improve the economic viability of active systems. In addition, combining intelligent and active packaging offers many intriguing possibilities, allowing for the development of more sophisticated packaging systems. Temperature indicators on packages will be used both to measure storage temperatures in the supply chain and as easy indicators to the consumer that a food product has been heated to a safe temperature.

Obviously, future trends are also prompted by commercial pressures, as manufacturers seek out ever more cost-effective solutions without compromising shelf life performance. In parallel, the requirement to meet environmental legislation, with particular regard to minimizing total packaging usage and substituting petrochemical materials with green polymers, also influences developments. The near future aims to expand eco-friendly solutions. Biodegradation of polymers offers an attractive route to environmentally sound waste management. Use of biopackaging will open up potential economic benefits to farmers and agricultural processors. Multicomponent films resembling synthetic packaging materials with excellent barrier and mechanical properties need to be developed. Cross-linking,

M.A. Del Nobile and A. Conte, *Packaging for Food Preservation*,
Food Engineering Series, DOI 10.1007/978-1-4614-7684-9,
© Springer Science+Business Media New York 2013

either chemically or enzymatically, of the various biomolecules is yet another approach of value in composite biodegradable films. Innovative techniques for preserving food safety and structural-nutritional integrity, as well as complete biodegradability, must be adopted. However, further scientific and technological developments are still needed to pave the way to the spreading of biodegradable or bio-based materials for food-packaging applications. A wider use of biopolymers in food-packaging applications, which would certainly lead to a lower environmental impact as compared to polymers from petrochemical sources, will be possible when problems related to the processability and performance of these materials are resolved.

If the overall environmental impact of a food-packaging system is to be reduced, then the function of the packaging should be included in the packaging design. Nowadays, packaging design has great significance for the success of food. The solution is in the implementation of preventive measures and, above all, in the adoption of policies shared by all actors in the production chain. Close cooperation between product and packaging experts is needed to develop guidelines for packaging companies. The aim is to create the optimal packaging system that satisfies all functional requirements in addition to meeting environmental and cost demands as much as possible. There is no doubt that the environmental impact can be significantly reduced if food losses decrease. It is important to develop the knowledge about the extent to which losses can be affected by packaging. It is still unclear, however, to what extent new packaging can influence food losses directly or indirectly by influencing consumer behavior. The main governmental goal in the European Union regarding packaging and the environment is to reduce the amount of packaging. Support for packaging development that reduces food losses is weak in the directive on packaging and packaging waste. If authorities want to reduce the total environmental impact of the food-packaging system, the text regarding the functions of packaging will have to become a higher priority in the packaging directive.

Index

M.A. Del Nobile and A. Conte, *Packaging for Food Preservation*,
Food Engineering Series, DOI 10.1007/978-1-4614-7684-9,
© Springer Science+Business Media New York 2013

Printed in the United States
By Bookmasters